Introduction to Molecular Medicine

Second Edition

Springer

New York
Berlin
Heidelberg
Barcelona
Budapest
Hong Kong
London
Milan
Paris
Santa Clara
Singapore
Tokyo

Dennis W. Ross

Introduction to
Molecular Medicine

Second Edition

With 60 Illustrations

Springer

Dennis W. Ross, M.D., Ph.D.
The School of Medicine
Department of Pathology
University of North Carolina at
  Chapel Hill
Chapel Hill, NC 27599, USA
and
Department of Pathology
Forsyth Memorial Hospital
3333 Silas Creek Parkway
Winston-Salem, NC 27104, USA

Library of Congress Cataloging-in-Publication Data
Ross, D. W. (Dennis W.)
    Introduction to molecular medicine / Dennis W. Ross. — 2nd ed.
        p.   cm.
    Includes bibliographical references and index.
    ISBN 0-387-94468-0 (softcover)
    1. Medical genetics.   2. Molecular biology.   3. Recombinant DNA.
4. Pathology, Molecular.   I. Title.
    [DNLM: 1. Genetics, Medical.   2. Molecular Biology.   3. Hereditary
Diseases — genetics.   4. Neoplasms — genetics.   QZ 50 R823i   1996]
RB155.R78   1996
616′.042 — dc20
DNLM/DLC   95-36653

Printed on acid-free paper.

Production managed by Terry Kornak; manufacturing supervised by Jeffrey Taub.
Typeset by Bytheway Typesetting Services, Inc., Norwich, NY.
Printed and bound by Braun-Brumfield, Inc., Ann Arbor, MI.
Printed in the United States of America.

9 8 7 6 5 4 3 2 1

ISBN 0-387-94468-0 Springer-Verlag New York Berlin Heidelberg   SPIN 10495100

Preface to the Second Edition

In the four years since the first edition of this book was published, the molecular revolution has continued. DNA has been named by *Time* magazine as the Molecule of the Year, a Nobel Prize has been awarded to a young man for the invention of the polymerase chain reaction, and television viewers have learned of the DNA fingerprint. Molecular technology in medicine is increasing. The availability of DNA probes for cancer susceptibility is stressing our system of insurance, testing our ideas about medical ethics, and teaching us new things about cancer. In this edition, I have added a number of new sections, as well as a new chapter. New examples of molecular medicine have been added to demonstrate current applications of this technology. The basic concepts of molecular biology remain the basis for the first three chapters of the book. The excitement surrounding molecular medicine that I mentioned in the preface to the first edition continues. It is now tinged with a touch of awe and a little bit of fear at the changes that recombinant DNA technology has brought to our society.

Preface to the First Edition

This book describes the discoveries that have created a field called molecular medicine. The use of recombinant DNA technology in medical research and most recently in medical practice constitutes a revolutionary tool in our study of disease. Probing the human genome is rapidly becoming as routine as looking at cells under a microscope. The cloning of a new gene is now a common occurrence, newspapers report. Recombinant DNA technology, like the invention of the microscope, shows us a world of detail richer than we might have imagined.

This book presents the discoveries, basic scientific concepts, and sense of excitement that surround the revolution in molecular medicine. The scientific basis of molecular medicine is explained in a simple and direct way. The level of technical detail, however, is sufficient for the reader to appreciate the power of recombinant DNA technology. This book is clinically oriented throughout. All of the examples and applications are related to medical discoveries and new methods of diagnosis and therapy. A few subjects within molecular medicine are examined in more detail to allow the reader to become aware of the strengths and shortcomings of a molecular approach to disease. I do not hide the incomplete understanding that still surrounds many of the recent discoveries in molecular medicine.

I intend to demonstrate the concepts of molecular medicine in this book by showing examples from all branches of medicine. I include, for instance, infectious diseases, genetic disorders, and cancer. However, I am not trying to be comprehensive in examining all areas of molecular medicine. So many discoveries are made each week in this field that it is not yet possible to draw them together in a comprehensive volume. My goal is to help the reader understand what the future may hold as well as the most important current applications.

This book is not a treatise, but an informal guide to a new field. As a guide, I try to communicate excitement, because this is the predominant feeling among people working in the field of molecular medicine.

Winston-Salem, NC   D.W.R.

Acknowledgments

Short summaries of some of the material in this book have appeared previously in my column "Advances in Science and Pathology," which appears in the *Archives of Pathology and Laboratory Medicine*. I would like to thank the editorial staff of the *Archives* and the American Medical Association for giving me the opportunity to develop my skills in science writing. A number of people have given me valuable suggestions after reviewing draft chapters of this book. I wish to thank them for their time and thought: Susan Atwater, Lanier Ayscue, Phil Carl, Joseph Dudley, Margaret Gulley, Charles Hassell, Edward Highsmith, Roy Hopfer, and William Kaufmann. I must also especially thank Daniel Sinclair, my research technician, for his considerable help on many features of the manuscript including preparing the computer graphics used as illustrations. I also want to thank Sarah Kielar, my secretary, for careful attention to the preparation of the typescript. I also wish to acknowledge most gratefully funding from The Blood Cell Fund, which has supported my work in cancer research and education.

Contents

II. Molecular Approach to Disease

PART I   BASICS OF MOLECULAR BIOLOGY

CHAPTER 1

Human Genome

The Genetic Message

The human genome consists of 6 billion nucleotides in a double-stranded helical deoxyribonucleic acid (DNA) molecule. This genetic message codes for the building and operation of the human body. The message is written in an alphabet that uses only four letters: A, C, G, and T. Each of the letters represents one of the four bases, which are the chemical building blocks of DNA: A—adenine, T—thymine, C—cytosine, and G—guanine. The nucleotides that spell out the genetic message are arrayed in a linear sequence in the double-stranded helical DNA molecule. The two strands of the DNA molecule are complementary copies of each other. The nucleotides on one strand pair with the complementary nucleotide on the other strand, as demonstrated in the top portion of Figure 1.1. A pairs with T; G pairs with C.

The genetic message is read not as single letters; rather, it is grouped into three-letter words. Each three-letter word is called a *codon*. A codon specifies one of the 20 possible amino acids that are the building blocks for all proteins. Figure 1.1 shows the organization of the genetic message at the level of DNA.

What has been called the fundamental paradigm of molecular biology is best stated as "one gene equals one enzyme." A gene encodes a specific linear sequence of amino acids assembled on the polyribosomes of the cell. The final form of the protein is achieved after spontaneous folding of the linear chain into a three-dimensional structure as demonstrated in Figure 1.2. It is this three-dimensional structure that gives the protein the ability to carry out its function. The sequence of amino acids in the linear chain predetermines the final folded structure of the protein. We have seen in Figure 1.1 that the nucleotide sequences in DNA serve as a template for messenger ribonucleic acid (mRNA), which directs the translation into a linear amino acid sequence. The last step, folding of the protein into a three-dimensional structure with specific enzymatic activity, is a spontaneous step. The protein structure is dependent on the gene, which is encoded

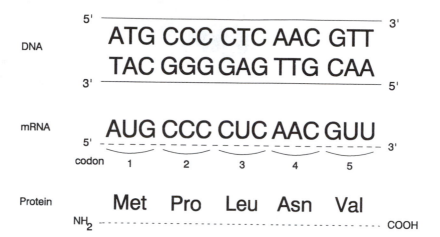

FIGURE 1.1. A genetic message begins as a double-stranded DNA molecule, which serves as the template for messenger RNA. The mRNA, in groups of three nucleotides to a codon, directs the order of amino acids in protein.

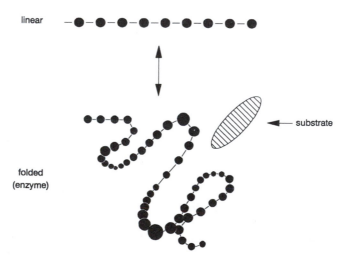

FIGURE 1.2. A chain of amino acids encoded by a gene and assembled on a polyribosome undergoes spontaneous folding to the final form of the protein. The form gives the protein its function, such as binding to a specific substrate in an enzyme-mediated reaction.

in the DNA. The events that control the expression of the genetic message control the function of the cells in the body by modulating the synthesis of proteins.

Shortly we will look at the details of how the genetic message specifies the synthesis of proteins and how proteins in turn regulate cell function. However, it is important first to get some idea of the amount of information contained in the human genome. A useful analogy compares the human genome to a library. The human genome is 6 billion letters long; there are only four letters in the genetic alphabet. The genome is written only with three-letter words. The message of the human genome is fragmented into 46 pieces, which we call chromosomes.

The library system of the University of North Carolina (UNC) has about the same number of letters in its library as the human genome—6 billion or a little bit more. The UNC library is written mostly in the English language, which has an alphabet of 26 letters plus ten numbers and some punctuation. The message in the library is organized into words, sentences, and books. The books are housed in 14 different library buildings around campus. We do not know the reason for the 14 different buildings nor do we know the reason for 46 human chromosomes. Maybe the number is not important.

We know a lot about the way information is organized in the library and used, and how it directs life on a university campus. Let's draw analogies to the human genome. To locate a book in the library, you go to the master card catalog (which is now computerized at UNC). You can look up a book by title, author, or subject. The card catalog tells you the Library of Congress call letters. A map of the library tells you where each call number section is physically located. For example, books about cancer with call letters "RC" are located on the eighth floor of the Davis Library. You can go the eighth floor and find the book on the shelf. If the book is a popular one, there may be several copies or several editions available. Taking the book you want down from the shelf, you use the index at the back of the book to find a particular subject of special interest. You might wish to photocopy a paragraph or two or to return to the circulation desk and check the book out. After you have consulted several books, you have the information you need. You may decide to write a report summarizing your research.

Using information from the human genome to synthesize a specific protein necessary for cell function is similar to consulting a library. We do not yet understand all the details of how the information is handled. To begin, some signal must tell the cell that it needs a certain protein. We call that event an inducer signal. Somehow the inducer signal causes the specific gene for the required protein to be located.

Finding the correct gene is like locating the necessary information in one paragraph of a specific book. We do not know how the cell does it. However, we do know that when the correct gene is found, a copy of the necessary information is transcribed into mRNA. This is the cell's method of photocopying information for use out of the library. The mRNA is then translated into protein. A number of genes may have to be transcribed to make the composite chains of the final protein. Thus the synthesis of the completed protein molecule is equivalent to our written report based on consulting several books.

Some major differences exist between the way we use a library and the way a cell uses the information stored in its genome; nevertheless the analogy is useful. The cell can find any gene it needs within seconds and synthesize a new protein within minutes. For a human to go to the library, research a subject, and write a report takes much longer. When a human writes a report, it is just possible that the report may find its way back into the library as a new book. This never happens with cells. The cell cannot add to its storehouse of information. The genome does not change rapidly, like a human library. The human genome changes by recombination of genes from parents as part of sexual reproduction, and by mutation, which is a rare event.

However, at the end of the 20th century, the rules for the human genome are changing. As you will see in reading this book, it is now possible via genetic engineering to write new information into the genome. Just as I am writing this book, I can also write and introduce new information into human cells. In the future, the human genome may become more like our libraries. For better or worse, the information written there will be, in part, the product of human ideas.

I will use the library analogy occasionally throughout this book. As we explore the applications of molecular biology to medicine, it is important not to lose track of the transfer of information, which is what DNA is all about.

The Genetic Code

Figure 1.3 shows the genetic code for translating the triplet nucleotide codon of mRNA into the amino acid in proteins. For mRNA, uracil (U) replaces thymidine (T), which is used only in DNA, not RNA. Since there are four nucleotides, read in groups of three, 64 possible combinations are available. The number of amino acids used as building blocks for proteins is only 21. This results in significant redundancy in the genetic code, with several different codons specifying the same amino acid. Examination of Figure 1.3 reveals that AUU, AUC, and AUA all code for isoleucine. The redundancy of the genetic code offers some protection against the adverse effects of mutation. If a single base-pair mutation occurs that alters the sequence codon from AUA to AUC, no change in the protein will occur. The genetic code is very old in evolution. Virtually all the living organisms on the planet use the code listed in Figure 1.3.

Anatomy of a Gene

Our knowledge of the structure of a gene, both physically in terms of its representation as a DNA molecule and functionally, is not yet complete,

| | | 2nd | | | |
	U	C	A	G	
U	Phe Phe Leu Leu	Ser Ser Ser Ser	Tyr Tyr STOP STOP	Cys Cys STOP Trp	U C A G
C	Leu Leu Leu Leu	Pro Pro Pro Pro	His His Gln Gln	Arg Arg Arg Arg	U C A G
A	Ile Ile Ile Met	Thr Thr Thr Thr	Asn Asn Lys Lys	Ser Ser Arg Arg	U C A G
G	Val Val Val Val	Ala Ala Ala Ala	Asp Asp Glu Glu	Gly Gly Gly Gly	U C A G

(1st axis on left, 3rd axis on right)

FIGURE 1.3. The genetic code for translation of the triplet nucleotide codons of mRNA into amino acids in proteins.

but much detail is known. A gene is a sequence of nucleotide base pairs that contains the genetic information necessary for directing synthesis of a protein. Figure 1.4 shows the schematic organization of a typical gene, in this instance the c-*myc* growth control oncogene. Figure 1.4 could be a page from an anatomy book of the future. Medical students of the 21st century will read not only gross anatomy and microscopic anatomy but a new text on the anatomy of the human genome.

A gene is organized into segments called *exons*, which are separated by *introns*. The exons contain the DNA sequences, which are transcribed into messenger RNA and then translated into proteins. The base pairs within the introns do not code for protein. There may be no major function associated with introns or the function may as yet may be undiscovered. In bacteria and other simple organisms, the gene sequences are continuous without introns. When a human gene with intron sequences is transcribed into mRNA, the entire DNA sequence from the start to the end of the gene is copied. Since the intron sequences do not contribute to the structure of the protein, these portions must be spliced out of the intermediate RNA transcript to form mRNA.

C-*myc*, as will be presented later in Chapter 8, is an oncogene on chromosome 8 that encodes for a nuclear binding protein that stimulates cell division. The inappropriate expression of c-*myc* protein, usually brought about

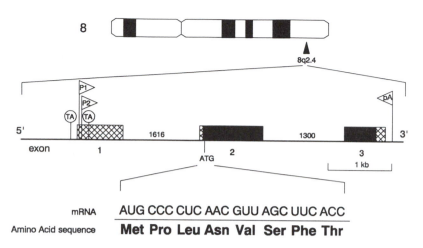

FIGURE 1.4. The anatomy of the c-*myc* oncogene shows its location on chromosome 8 (arrowhead) and its grouping into three exons. The initiation sequence ATG starts the mRNA template coding for the amino acid sequence of the final protein product.

by structural aberrations in the gene's anatomy, is associated with a number of human malignancies. A good example is Burkitt's lymphoma, in which a t(8;14) chromosomal translocation erroneously juxtaposes c-*myc* with the immunoglobulin gene from chromosome 14.

At the top of Figure 1.4 is a schematic diagram of chromosome 8 as it would appear in a Giemsa-stained metaphase karyotype. The light- and dark-staining bands on the short (p) and the long (q) arms of the chromosome define physical areas on the chromosome. The location of c-*myc* on the long arm at 8q2.4 is indicated by an arrowhead.

The center panel of Figure 1.4 shows the genomic structure of c-*myc*. C-*myc* is a small gene, consisting of three exons and spanning only 5,000 base pairs. This is a distance of 0.005 *centiMorgans*. A centiMorgan is defined as a distance along a chromosome that has a 1% probability of undergoing genetic recombination during gamete formation at meiosis. (See also Chapter 6.) As genes get further apart along a chromosome, there is an increasing (though still slight) chance that they will sort independently (as if on separate chromosomes) during sexual reproduction. Genes that are on different chromosomes always sort independently.

The spacing between exons 1 and 2 of the c-*myc* gene is 1,616 base pairs, and between exons 2 and 3 the spacing is about 1,300 base pairs. Important structures within the *myc* gene are indicated in Figure 1.4. P1 and P2 are two promoters that help control the expression of this gene. A promoter is a region of DNA associated with a gene that regulates gene expression. There are other regulatory elements associated with genes besides promoters. For example, the c-*myc* gene has two TATAA boxes, near exon 1 (marked "TA" in Fig. 1.4). TATAA boxes are also regulators of gene initiation. C-*myc* is unusual as a gene in that exon 1 contains no codons for the c-*myc* protein, but exon 1 apparently has a regulatory function. Coding for the protein begins with the ATG sequence 16 bp (bp = base pairs) from the start of exon 2 and continues through the second and most of the third exon. At the end of the third exon are coding sequences for a poly A tail common to the end of most messenger RNA molecules. The intron sequences between exons 2 and 3 are initially transcribed into RNA but are then spliced out before the final messenger RNA molecule is exported to the cytoplasm. The c-*myc* messenger RNA is approximately 2,300 bp long and codes for a protein of 439 amino acids.

The entire base-pair sequence for c-*myc* is known for humans and several nonhuman species. There are very few differences in the genomic structure of c-*myc* between species. This is the case in general for oncogenes that are highly conserved in terms of their structure throughout evolution. The complete DNA sequence for the second exon of the c-*myc* oncogene is shown in Figure 1.5. Let us look at some of the detailed structure of the gene as represented in this sequence data. I have annotated the sequence data at important sites. Point 1 is near the end of the intron between exons 1 and 2. Point 2 denotes a box that encompasses all of the exon 2 sequences. Despite the fact that exon 2 starts at point 2 with the codon CAG, it is not until point 3 that RNA translation into protein is initiated. The codon ATG at point 3 is the initiation codon for RNA to protein translation. The ATG initiation codon is also marked on Figure 1.4 as an important element in the overall anatomy of the c-*myc* gene. The ATG initiation triplet always codes for the amino acid methionine. From point 3 until the end of exon 2, the amino acid for each codon is written above the genomic sequence. The second exon of c-*myc* terminates with the codon TCT, which codes for the amino acid serine, and is marked at point 4. Point 5 denotes the start of the second intron. The second intron spans more than 1,300 base pairs before exon 3 is reached.

The DNA sequence data given in Figure 1.5 for the c-*myc* oncogene shows some of the challenges facing the human genome project or facing any lab dealing with DNA sequencing data. The raw data for DNA sequencing are derived from an autoradiograph of a sequencing gel. An example is shown in Figure 1.6. Each band corresponds to one nucleotide in the DNA molecule. The four columns correspond to the four nucleotides A, C, G, and T. The gel is read from bottom to top. At each step on the ladder progressing from bottom to top, the investigator determines whether the rung on the ladder is in the A, C, G, or T column. As the gel is read, each band is decoded into one of the four letters. A typical sequencing gel such as shown in this figure can decode 200 to 500 base pairs.

As DNA sequence data are acquired, they must be analyzed. Until the entire sequence for a gene is assembled, we do not know where a particular

2

3 Met Pro Leu Asn Val Ser Phe Thr Asn Arg Asn Tyr Asp Leu Asp Tyr Asp Ser Val

1 CCGCTCCAGCAGCCTCCCGCGACG|ATG|CCCCTCAACGTTAGCTTCACCAACAGGAACTATGACCTCGACTACGACTCGGT

Gln Pro Tyr Phe Tyr Cys Asp Glu Glu Glu Asn Phe Tyr His Gln Gln Gln Ser Glu Leu Gln Pro Pro Ala Pro
GCAGCCGTATTTCTACTGCGACGAGGAGGAGAACTTCTACCAGCAGCAGCAGAGCGAGCTGCAGCCCCCGGCGCCC

Ser Glu Asp Ile Trp Lys Lys Phe Glu Leu Leu Pro Thr Pro Pro Leu Ser Pro Ser Arg Arg Ser Gly Leu Cys Ser Pro
AGCGAGGAGGATATCTGGAAGAAATTCGAGCTGCTGCCCACCCCGCCCCTGTCCCCTAGCCGGCGCTCCGGGCTCTGCTCGCCC

Ser Tyr Val Ala Val Ala Thr Pro Phe Ser Leu Arg Gly Asp Asn Asp Gly Gly Gly Gly Ser Phe Ser Thr Ala Asp Gln
TCCTACGTTGCGGTCGCTACACCCTTCTCCCTTCGGGGAGACAACGACGGCGGTGGCGGGAGCTTCTCCACGGCCGACCAG

Leu Glu Met Val Thr Glu Leu Leu Gly Gly Asp Met Val Asn Gln Ser Phe Ile Cys Asp Pro Asp Asp Glu Thr Phe
CTGGAGATGGTGACCGAGCTGCTGGGAGGAGACATGGTGAACCAGAGTTTCATCTGCGACCCGGACGACGAGACCTT

Ile Lys Asn Ile Ile Gln Asp Cys Met Trp Ser Gly Phe Ser Ala Ala Ala Lys Leu Val Ser Glu Lys Leu Ala
CATCAAAAACATCATCCAGGACTGTATGTGGAGCGGCTTCTCGGCCGCCGCCAAGCTCGTTCAGAGAAGCTGGCC

Ser Tyr Gln Ala Ala Arg Lys Asp Ser Gly Ser Pro Asn Pro Ala Arg Gly His Ser Val Cys Ser Thr Ser Ser Leu
TCCTACCAGGCTGCGCGCAAAGACAGCGGCAGCCCGAACCCCGCGCCGGCCAGCGGTCGTGCTCGCCACCTCCAGCTTGTA

Tyr Leu Gln Asp Leu Ser Ala Ala Ala Ser Glu Cys Ile Asp Pro Ser Val Val Phe Pro Tyr Pro Leu Asn Asp Ser
CCTGCAGGATCTGAGCGCCGCGCCTCAGAGTGCATCGACCCCTCGGTGGTCTTCCCCTACCCTCTCAACGACAGCAGCT

Ser Ser Pro Lys Ser Cys Ala Ser Gln Asp Ser Ser Ala Phe Ser Pro Ser Ser Asp Ser Leu Leu Ser Ser Thr Glu Ser Ser
CGCCCAAGTCCTGCGCCTCGCAAGACTCCAGCGCCTTCTCTCCGTCCTCCGGATTCTCTGCTCTCCTCGACGGAGTCCTCC

Pro Gln Gly Ser Pro Glu Pro Leu Val Leu His Glu Glu Thr Pro Pro Thr Thr Ser Ser Asp Ser
CCGCAGGGCAGCCCCGAGCCCCTGGTGCTCCATGAGGAGACACCGCCCACCACCAGCAGCGACTCTGG|TAAGCGAAGC

4   5

FIGURE 1.5. The nucleotide sequence for exon 2 of the c-*myc* oncogene is given with the corresponding amino acids.

FIGURE 1.6. An autoradiograph of a DNA sequencing gel, such as shown here, can resolve each individual nucleotide within a short segment of DNA. Each of the four lanes corresponds to one of the four nucleotides: A, C, G, T. (Autoradiograph provided by Georgette Dent, MD, Department of Pathology, University of North Carolina at Chapel Hill.)

stretch of DNA base pairs "fits in." To analyze DNA sequence data, we first look for an *open reading frame*. An open reading frame is a series of base pairs in which the stop codons are not encountered for a long stretch. We know that DNA sequences are read three letters at a time by the cell. Three codons (as indicated in Fig. 1.3)—UAG, UAA, and UGA—are called stop codons, because they specify a stop to the translation of RNA into protein. A stop codon is like a period at the end of a sentence. In searching for an open reading frame, we have an additional problem in that we do not have a frame of reference in which to begin. We do not know at which letter to start, since we are in the middle of the message. To search for an open reading frame, we try all three of the possible reading frames. We begin at the first base pair and read in groups of three, then shift to the second base pair and read in groups of three, and finally shift to the third base pair. Consider the following example, written in three-letter English words instead of three-nucleotide codons:

FTHEOWLANDCATATECODONEANDALLEG.

An open reading frame that produces a sensible message begins at the second letter. Reading frames starting at the first or third letter produce messages that make no sense.

Finding the correct reading frame can be difficult. Only when one of our three possible reading frames reveals a long "open" stretch without stop codons can we be sure that we have found it. The theory is that only within a gene, and only if the correct frame of reference is used, will a long stretch of DNA sequences be found in which none of the three stop codons occurs. Everywhere else (incorrect frames of reference and in parts of the DNA that are outside of gene sequences), the stop codons will be encountered by chance every 30 or so bases. If all of the DNA sequences we are currently analyzing are from an intron, or from noncoding DNA between genes, then it will not be possible to find the correct reading frame. As you can imagine, laboratory workers who have been slowly sequencing an unknown stretch of DNA for several months can get quite excited when they finally find an open reading frame.

The decoding analysis is tedious. Fortunately, computer programs have greatly simplified the task. The sequences shown in Figure 1.5, when analyzed, show that an open reading frame begins at point 2 with the codon CAG. From there, the open reading frame continues for a stretch of 765 base pairs until the stop codon TAA (equivalent to UAA when transcribed into mRNA) is reached.

In addition to looking for the structure of a gene by searching for open reading frames, we also look for other structures that will be described later such as TATAA boxes, restriction enzyme cut sites, or special structures such as long terminal repeats (LTR). All of this detailed anatomy of a gene can be ascertained in part by looking at the raw sequence data. Once the structural anatomy of a gene is known, the next challenge is to understand its function. What protein does the gene make? What is that protein's function? How is the expression of the gene regulated? These questions will be discussed in Chapter 2, but first we need to know how DNA is replicated and a bit more about the human genome project, which seeks to decode the entire genome.

Physical Organization of the Genome

The picture of the gene that I have given so far is a functional one. Figure 1.4 is a schematic of the various elements of a gene. This functional anatomy of the gene, with exons, introns, and promoters, helps us understand how a gene works, but it does not tell us what a gene looks like. Even Figure 1.5, which gives the entire base-pair sequence of a short portion of a gene, is not at all a physical picture. If you were to look at a piece of DNA, these functional structures would not be visible.

The physical organization of the genome, though very different from the functional anatomy, is also important. The nucleus of a cell is only a couple of microns across and we have to put a couple meters of a linear DNA molecule into it. That is like putting ten thousand miles of string into an average-sized room. It is not easy. Furthermore, you have to be able to find one little part on that string in seconds.

The physical organization of DNA is accomplished through a successive hierarchy of packing structures, as represented in Figure 1.7. The linear double-helix DNA molecule is wrapped for two turns around small, globular proteins called histones. These histone beads are grouped into a winding helical secondary structure of their own. Then the secondary structure is packed into loops. When a cell undergoes mitosis, these loops gather together in a helical structure that makes up the arms of the visible chromosome. A picture of even a very small portion of DNA would appear as an incredible tangle of a long, thin molecule wrapped around the histones, like thread on bobbins. So there is really a tremendous hierarchy of physical structure to DNA—at least four levels of increasingly complex structure. We do not understand at all how a small portion of the genome can be accessed quickly. How can a single piece of DNA be located, opened up, read, and closed? It is important to realize what we do not know as well as what we do know.

Gene Maps

There are two different ways to express how DNA is organized: physical and genetic maps. Figure 1.4 is an example of a physical map. The c-*myc* gene is localized to a place on the long arm of chromosome 8 (8q2.4 in chromosome banding nomenclature). Figure 5.3 in Chapter 5 is an example of a genetic map (of the HIV-1 virus) that shows the spatial relationship of the genes composing the viral genome of HIV. Shading on the map gives some indication of how the various genes and regulatory sequences interact. Both physical and genetic maps are common means for presenting genetic

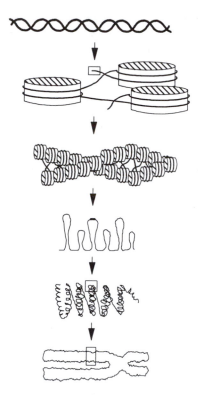

FIGURE 1.7. A schematic representation of the physical structure of a gene shows multiple levels of packing of the DNA molecule into the final chromosome structure.

data. When the human genome is more completely understood, we will have physical as well as genetic maps for all of it. The physical map will show us where genes are on the various chromosomes and in relation to other genes. The genetic map will indicate hot spots for regulatory sequences, mutations, recombinations, and other events. A physical map is close to our current concept of a road map; a genetic map is more like a program detailing the timing of local events.

Human Genome Project

One of the largest-scale biomedical research projects ever undertaken is now under way—the human genome project (Cantor, 1990; Watson, 1990). The goal is to sequence the 6 billion base pairs that constitute the human genome. The task is a formidable one. Sequencing methods currently use thin-layer vertical gel electrophoresis followed by autoradiographs of gel blots, as shown in Figure 1.6. This is a very labor-intensive technology. A single sequencing gel can decode several hundred base pairs. Without any duplication of effort, or repeat testing, this approach would require on the order of tens of millions of gels to encompass the entire human genome!

An analogy can be made between the human genome project and an immense jigsaw puzzle with millions of individual pieces. Each piece represents a cloned segment of the human genome contained in a plasmid. (Cloning will be described in Chapter 3.) At the start, each laboratory begins decoding the letters on one piece. Next the investigator wants to join his lab's piece into its proper place in the puzzle. Early on in the human genome project this will be difficult. It is likely that no other pieces nearby have been decoded. The puzzle begins to come together in little clusters far separated from each other. Part of the organization of the human genome project has been designed to establish signposts scattered throughout the genome. These signposts will allow individual investigators to know where in the entire puzzle they are working. Each lab can speak to other laboratories in terms of distance from the signposts. An individual laboratory can transmit its newly discovered sequences to the human genome data bank computer. The data become available to all subscribers immediately. Each laboratory, as it adds new sequences, can see if any other lab has come up on this section of the puzzle.

Analysis of nucleotide sequence data as they are being generated can speed the process of decoding the human genome. When computer analysis finds an open reading frame indicating the presence of a gene, the probable amino acid sequence for the gene's protein can also be determined. The amino acid sequence often gives some suggestions as to the function of the protein product of the gene. Once enough of the new gene sequence has been determined, the amino acid structure can be guessed. A protein with a hydrophobic tail would probably be found bound to a membrane. Other portions of the protein may show homology—for example, to a kinase, which is an enzyme affecting energy and transport pathways in the cell. The combination of the hydrophobic tail with an active kinase region would suggest that the newly discovered gene is a surface receptor protein such as a tyrosine kinase.

Many controversies surround setting priorities for the human genome project. With manual sequencing technology as it now exists, the primary limit to the rapidity of sequencing the human genome would be the rate of funding. However, some investigators feel that if more time is spent at the beginning of the project to engineer high-speed automated sequencing technology, the end result will be achieved not only sooner, but at much less expense. Laboratories currently setting out to sequence the genome at the rate of 100 base pairs a day might at the end of the year have some 30,000 base pairs decoded. A neighboring laboratory that bets its research monies on advanced technology might not have even started after 1 year. However, when one of the automated labs starts it might be able to generate 30,000 base-pair sequences per day. Once completed, the human genome will occupy some 10,000 conventional floppy disks when spelled out.

The scientific and medical benefits of sequencing the human genome are quite significant and many scientists and physicians feel that the potential benefits are appropriate to the very large cost. Sequencing the human genome will make diagnostic probes available for many diseases. The human genome project by virtue of its immense size is not only likely to introduce a great deal of new information to the understanding of human disease, but will also create a new scientific culture. The goal of the human genome project is immense in its total scope, but each individual laboratory's contribution becomes blurred, and in fact submerged, in the common goal. Individual investigators may submit large amounts of sequence data that they themselves do not understand and cannot interpret until the data have been joined up with the entire body of DNA sequencing results.

Bibliography

Alberts B, Bray D, Lewis J, Raff M, Roberts K, Watson JD (eds) (1994) *Molecular Biology of the Cell*. Garland Publishing Inc, NY.

Cantor CR (1990) Orchestrating the human genome project. Science 248:49–51.

Darnell J, Lodish H, Baltimore D (eds) (1990) *Molecular Cell Biology*. WH Freeman Press, NY.

Lewin B (1994) *Genes V*. Oxford University Press, NY.

Watson JD (1990) The human genome project: past, present and future. Science 248:44–48.

Watson JD, Gilman M, Witkowski J, Zoller M (1992) *Recombinant DNA*, 2nd edition. WH Freeman, NY.

CHAPTER 2

Gene Regulation and Expression

Overview

A gene is a segment of DNA consisting of codons specifying the amino acids for a protein and of control sequences that regulate gene expression. When a cell divides, the entire human genome, that is, all the DNA stored within the nucleus, must be copied. In Figure 2.1 the replication of DNA is demonstrated. The regulation of a cell's function is controlled by altering gene expression. The steps in gene expression are (1) transcription of the gene's DNA to RNA, (2) RNA processing to produce a spliced mRNA, (3) translation of mRNA on a polyribosome to a polypeptide chain, and (4) final protein processing to the functional tertiary form. Figures 2.2 through 2.5 illustrate the steps in this process, which will be covered in more detail later in this chapter.

DNA Replication

Every time a cell divides, it must first make a duplicate copy of its genome to share with each of its two progeny. As an embryo grows from a single cell to a multicellular organism, its genomic information must be faithfully reproduced in every cell. A human cell typically takes 6 to 12 hours to duplicate its DNA. At the completion of this period of DNA synthesis, called the S phase of the cell division cycle, the cell is tetraploid and contains 92 chromosomes. This tetraploid state persists for several more hours and then the cell divides by mitosis. Each of the two new cells receives identical copies of the 46-chromosome diploid genome.

The process of DNA duplication is complex in all of its detail, yet the overall plan is simple. Figure 2.1 is a schematic representation of the major events. The process of DNA replication is dependent on specialized enzymes. These enzymes are also the key tools of recombinant DNA technology.

The replication of DNA begins with a separation of the double-stranded DNA helix. This process is called "melting" because it takes the equivalent

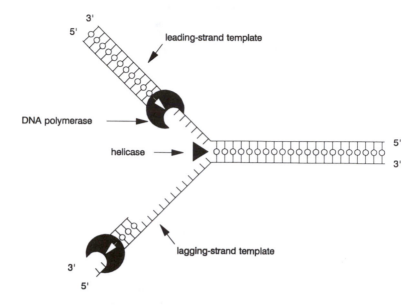

FIGURE 2.1. The process of DNA replication is shown in a simplified schematic form.

of thermal energy to occur. Heating double-stranded DNA in vitro to 90°C will cause it to separate into single-strand components. In the cell, DNA strand separation begins at multiple spots throughout the genome. Enzymes catalyze the process of strand separation, allowing the DNA to "melt" at 37°C. Each local area where strand separation occurs is the start of a

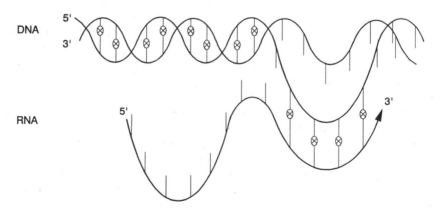

FIGURE 2.2. The double-stranded DNA template partially unwinds, allowing an RNA transcript to be made.

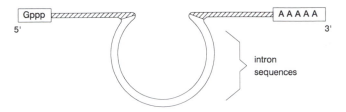

FIGURE 2.3. The RNA transcript molecule is processed to remove intron sequences that do not contribute to the coding of a protein. A cap is added to the 5′ end and a tail of adenine molecules is added at the 3′ end.

structure called a *replicon*. Several enzymes are involved in initiating and controlling this process. An enzyme called *DNA helicase* controls the unwinding of the double-stranded DNA molecule. DNA polymerase catalyzes the synthesis of a new duplicate strand. Deoxyribonucleic acid polymerase works by adding nucleotides at the 3′ end of the growing newly synthesized strand. The nucleotides are added by copying the template strand by using complementary bases. One of the two original DNA strands is directly copied in the 5′ to 3′ direction. The other strand must be duplicated in the reverse 3′ to 5′ direction. The reverse strand duplication occurs as a synthesis of small fragments made in a 5′ to 3′ direction (the only direction that DNA polymerase works in). Then these small fragments are joined head to tail. Figure 2.1 demonstrates this process by showing a single DNA replicon. As the DNA replication extends in both directions, the newly synthesized strands re-form a double helix with each of the original template strands. This is called semiconservative replication. Each copy of the DNA will contain one original strand from the parent molecule and one newly synthesized strand twisted together in a double helix. A famous experiment

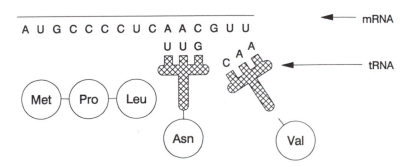

FIGURE 2.4. The mRNA is translated into protein, one codon at a time, by tRNA molecules that recognize the codon and transport the appropriate amino acid to the growing peptide chain.

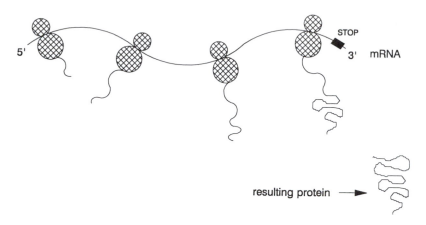

FIGURE 2.5. Protein synthesis from the mRNA template occurs via the machinery of polyribosomes in the cell cytoplasm. Ribosomes move along the mRNA until a stop codon is reached. At that point the ribosome falls off the mRNA and releases its peptide chain, which now folds into the final form of the protein molecule, which is the end product of the gene.

in 1958 by Meselson and Stahl that used chemical tracers showed that the genetic material of bacteria was replicated in this semiconservative fashion. This was the first experiment whose results gave insight into the replication process of the genome.

If we return to our library analogy for the human genome, it is easy to appreciate the problem of duplicating a million or so books. In the case of the human genome, we must make an exact copy of 6 billion base pairs within the space of 6 to 12 hours. Not only must the process be very fast, but it must be incredibly precise. Any error would be a permanent mutation and persist as a change in the genome. The original single DNA molecule of the genome present in the fertilized zygote at the moment of conception must be copied some 10^{15} times during the subsequent course of a human lifetime. Some of these 10^{15} divisions occur as the body grows from a single cell to an adult mass of 70 kg of cells. Many more cell divisions occur throughout our life span as the cells in our body continually renew themselves. Even a small error rate in DNA replication would produce a tremendous number of changes in the genome.

The precision of DNA replication is maintained by several systems. First, the DNA-copying apparatus of the replicon makes few errors. The measured error rate in direct copying is 0.01%, or one incorrect base in every 10^4 base pairs. Several error-correction systems, however, make the final error rate much lower. There is a proofreading system that checks the tentative match between the template strand and the newly synthesized duplicate strand. Other systems detect chemical modification such as methylation of bases. Altogether these quality-assurance systems greatly improve the error rate in DNA replication. The true precision of DNA replication with all error-correcting systems in place is estimated to be one in 10^8 base pairs. This still means that some 60 errors occur every time the human genome is copied as part of cell division.

Fortunately, most errors occur just by chance in spaces outside of genes. Even errors occurring within genes have only a small likelihood of affecting the function of the gene's protein. Furthermore, since any one cell utilizes only a relatively small number of its genes, the likelihood of an error in a critical gene necessary for the function of the cell is very low. To again use the library analogy, if only 60 misspelled words occur throughout the entire library, the chance of one reader encountering an error and being misled would be very slight.

Gene Expression

When a gene is turned on, its genetic sequences are translated into a protein product. A primary mechanism for activating a gene is the binding of an inducer protein to a promoter site at the head of the gene. The gene remains in a turned-on state, directing the synthesis of more protein, until something causes it to be switched off. A common mechanism of gene regulation is negative feedback. The presence of a large amount of the protein product from the gene can interfere with the binding of the inducer that originally turned the gene on. When this happens, the gene is turned off. This thermostatlike mechanism turns genes on and off, maintaining a supply of product balanced against consumption.

A mechanism used to regulate gene expression is external signaling. The way external signals influence gene expression is incompletely understood; further details are emerging as research progresses into this area. We do know that cell-surface receptors have the ability to recognize and bind specific molecules. When a specific external-signal molecule binds to the cell receptor, an enzyme called protein kinase located on the internal side of the cell membrane is activated. The protein kinase travels through the cell cytoplasm and into the cell nucleus. There, through other intermediate mechanisms including binding of an inducer protein to a promoter sequence, the target gene is turned on or off. The number of receptors regulates the speed and intensity of the response to an outside signal. Cell-surface receptors are themselves under gene control. The number of receptors can be increased or decreased. If the expression of a cell-surface receptor gene is down-regulated, eventually all cell-surface receptors will be lost. When the cell has no surface receptors, it will no longer respond to the external signal of that receptor. The loss of cell-surface receptors is a means of putting a cell in a latent state with regard to reacting to outside stimuli. The topic of external signaling as a means of control of gene expression will be considered again in Chapter 8, whose topic is cancer. Figure 8.1 schematically demonstrates the signal pathway.

Transcription

The first step in gene expression after an inducer protein binds to the promoter region just 5′ to the first exon is transcription. As demonstrated in Figure 2.2, the DNA begins to unwind at the 5′end of the gene. A single-stranded RNA molecule is made from the DNA template. This RNA represents an exact copy of the anticoding DNA strand, which is also called the complementary strand of the DNA molecule. One important difference compared to DNA is that RNA uses the base called uracil to replace thymidine. (All T's are replaced by U's in the RNA molecule copy.) In Figure 1.4, which displays the overall process of gene expression, I showed the DNA sequence on the coding strand and the mRNA sequence that matches it. The convention among molecular biologists is to list the sequences as they occur on the coding strand of DNA that match the mRNA sequences synthesized from the complementary DNA strand.

After transcription in the nucleus, the RNA molecule moves out into the cytoplasm, where it is processed to remove the portions that correspond to intron sequences. This is shown schematically in Figure 2.3. The portions of the RNA that correspond to intron sequences are spliced out. The start and end of each exon have sequences called splice acceptor and donor sites. These sites, when copied onto the RNA molecule, signal an enzyme to remove the intervening parts that are not needed to code for the structure of the protein. A final step in RNA processing is to add a cap to the head and a series of adenines to the tail of the molecule. The completed structure is denoted mRNA.

Translation

The mRNA now binds to a *polyribosome*, a cytoplasmic organelle that serves as the site where mRNA is translated into protein. The polyribosome moves along the mRNA molecule. Each group of three nucleotides on the mRNA molecule represents one codon. Each codon attracts a carrier molecule called transfer RNA (tRNA). The tRNA on one end matches the codon on the mRNA. On the other end of the tRNA is attached a single amino acid. There is a specific tRNA carrier for each codon, and that tRNA can transport only one kind of amino acid. The tRNAs carry the amino acids to the vicinity of the mRNA template and attach them like beads on a string, one at a time. Figure 2.4 demonstrates the building up of a peptide, one amino acid at a time, as transfer RNA molecules read the template mRNA.

This translation of mRNA to protein continues until a stop codon at the end of the mRNA molecule is reached. When the stop codon is reached, the mRNA falls off the polyribosome, as is shown in Figure 2.5. The used mRNA very quickly degrades. The completed peptide falls off the other end of the polyribosome. The peptide now undergoes folding into the secondary and tertiary structure of the mature protein (as shown in Fig. 1.2). For some proteins, additional modifications are carried out by the cell, such as adding sugar groups to make glycosylated proteins. The protein is now ready for use in the cell.

Regulation

The overall process of regulating gene function is complex, with rich detail and many as-yet-undiscovered nuances. The orchestration of gene expression occurs at many levels and allows for a great deal of control over gene function. We will consider some artificial tools for regulating gene function in the next two sections — antisense oligonucleotides and viral vectors — but first let us summarize the natural steps by which a cell modulates gene expression. At the topmost level, even before being read into RNA, genes can exist in "active" or "inactive" forms. Some gene systems consist of multiple pieces that must be joined together before transcription. In Chapter 7 we will see how this is the case for the genes of the immune system.

Much of the control of gene function occurs at regulating how a gene is transcribed into RNA. Our page from the future gene anatomy textbook, Figure 1.4, shows the c-*myc* oncogene. There are two promoter sites, P1 and P2, shown just to the 5′ direction of the first exon. Two TATAA sites are nearby. Promoters and TATAA sites are necessary to facilitate binding of RNA polymerase at the head of the gene. Thus they can be used to control the rate of transcription. Other factors, such as enhancers, are located further away from the target gene and fall outside of the gene map shown in Figure 1.4. Enhancers also regulate transcription. Other mechanisms, including frame shift modifications and factors that influence protein binding to DNA, also regulate RNA transcription. The overall control of transcription is complex, allowing for multiple factors to influence the activity of a gene.

There are perhaps fewer factors controlling the next step in gene expression, translation of mRNA into protein, or maybe these factors have not yet been discovered. The polyribosome factory that runs along the mRNA strand as shown in Figure 2.5 is very efficient at producing linear strands of amino acid sequences matching the genome blueprint. The last step in producing the final gene product is folding the protein into the active form and adding any posttranslational touches such as glycosylation.

Antisense Oligonucleotides

Genes can be regulated at the level of translation by a new class of drugs called antisense oligonucleotides. Antisense oligonucleotides constitute a means of interfering with the binding of mRNA to the polyribosome. A short piece of DNA, typically 15 to 30 base pairs long, is synthesized. This oligonucleotide is complementary, a genetic mirror image, to a portion of the much longer mRNA molecule. This is called an antisense copy since it is complementary to the mRNA that contains the genetic message in the correct "sense" orientation. The antisense oligonucleotide binds to its complementary portion of the mRNA molecule and produces a short double-stranded sequence along this otherwise-linear single-stranded molecule. This double-stranded area prevents the mRNA from being translated into proteins on the polyribosomes. Furthermore, double-stranded portions of mRNA are recognized as abnormal by the cell and are destroyed by an enzyme called ribonuclease H. Thus when an antisense oligonucleotide binds to its complementary region on an mRNA molecule, the result is the functional destruction of that message. If the antisense oligonucleotide is present in excess in the cytoplasm, no message can be successfully processed by the cell. Thus, despite the fact that a specific gene is turned on and actively being transcribed into mRNA, the message cannot get through. Antisense oligonucleotides constitute a very specific mechanism for interfering with the expression of only a single gene.

There are many applications of antisense oligonucleotide suppression of gene expression. One such application is to shut off certain genes whose expression is not desirable. Addition of antisense oligonucleotide specific to the c-*myc* oncogene prevents expression of this gene in tumors, resulting in cessation of cell division. Antisense c-*myc* oligonucleotides have been shown in cell culture models to be a specific way of arresting cell growth.

Antisense oligonucleotides constitute a very powerful research tool that can be used to understand the effects of a newly discovered gene. With antisense oligonucleotides, a single gene can be shut off and the resulting effect on cell or tissue function can be observed. This technique has been applied to the study of numerous oncogenes, embryonic development genes (such as the wingless gene in *Drosophila*), and structural protein genes such as actin and myosin.

Some important technical problems remain as a barrier to the expanded use of antisense oligonucleotides for medical therapy. Oligonucleotides are intrinsically unstable, being rapidly degraded by enzymes within the cell and in the blood. However, synthesis of new forms of antisense oligonucleotides with a modified chemical backbone offers increased stability and overcomes the normally rapid degradation of these molecules. Additionally, antisense oligonucleotides are relatively large molecules with molecular weight in the 5- to 15-kd range. This limits their ability to penetrate into the cell cytoplasm. Nevertheless, the use of antisense oligonucleotides both for research, as antiviral agents, and as specific regulators of gene expression to treat human disease, remains another important new tool derived from recombinant DNA technology.

Viral Vectors

Gene expression can also be altered in human cells by introducing a virus that has been engineered to carry a cloned fragment of DNA into the cell. The virus is called a vector and the process is transfection. An engineered virus could, for example, infect the cell and continuously produce antisense copies of mRNA. This would block permanently the gene specified by the antisense fragment. Many viruses are available as vectors, each with special properties that assist in the "engineering" of controlled gene expression. Viruses are infective, which makes them easier to introduce into cells compared with bare pieces of DNA. Some viruses only infect certain cell types, which gives another parameter for controlling the process. The special class of RNA retroviruses insert their message permanently into the nuclear DNA of the cell. This alters the genome of that cell, and of all future progeny of that cell, making the organism "transgenic."

Table 2.1 lists some common examples of vectors used for cloning gene fragments (see also Chapter 3) and for transfection. These vectors are similar to computer disks. They are the means by which scientists "carry" pieces of DNA. For example, I might request a collaborator to send me a fragment of the *bcr* gene (involved in chronic myeloid leukemia; see Chapter 8). Just before hanging up the phone, I hope I would remember to ask, "What's the vector?" If the response is "pBR322," I know that I can easily handle this standard plasmid. I can grow pBR322 in a culture of *Escherichia coli* and make millions of copies of the *bcr* piece inserted into the vector. Like computer disks, vectors are constantly being upgraded with better features and more carrying capacity. For vectors we talk of storage capacity in kilobases instead of megabytes as on computer disks.

An example of altering gene expression with a vector is the treatment of cystic fibrosis (see also discussion in Chapter 6). An adenovirus that has been engineered to carry a cloned fragment of the normal human cystic fibrosis transmembrane conductance regulator gene (CFTR) is allowed to infect the lung cells of CF patients. The lung cells now have a working copy of the gene for chloride ion channel conductance and the abnormally thick bronchial secretions characteristic of CF are reduced as this gene is expressed. An animal model of CF has demonstrated the feasibility of this form of gene therapy and clinical trials have begun.

Table 2.1. Examples of vectors.

Name	Origin	Length	Example	Growth
Plasmid	Natural DNA fragment in bacteria	10^3 bp	pBR322	*E. coli*
Phage	Bacterial virus	10^4 bp	λgt10	*E. coli*
Cosmid	Bioengineered	10^4 bp	c2RB	*E. coli*
YAC	Yeast artificial chromosome	10^5 bp	pJS97/98	Yeast
Adenovirus	Human virus	10^5 bp	Ad-2	Human cells

Bibliography

Culotta E, Koshland DE Jr (1992) No news is good news. Science 258:1862–1865.

Stein CA, Cheng YC (1993) Antisense oligonucleotides as therapeutic agents — is the bullet really magical? Science 261:1004–1012.

Stein CA, Cohen JS (1989) Oligodeoxynucleotides as inhibitors of gene expression: a review. Cancer Res 48:2659–2668.

van der Krol AR, Mol JNM, Stuitje AR (1988) Modulation of eukaryotic gene expression by complementary RNA or DNA sequences. Biotechniques 6:958–976.

CHAPTER 3

Tools of Recombinant DNA Technology

Restriction Enzymes

Restriction endonucleases, also called restriction enzymes, are bacterial proteins that cut the long, linear DNA molecule into fragments. Restriction enzymes are a major tool of recombinant DNA technology. A restriction enzyme recognizes a specific nucleotide sequence, such as AGCT, and cuts DNA wherever that combination of "letters" occurs. These enzymes are isolated from bacteria and named with a three- or four-letter sequence followed by a roman numeral. For example, EcoRI is a restriction enzyme isolated from *Escherichia coli*.

It seems peculiar that an enzyme such as EcoRI would exist. Why would bacteria produce an enzyme that cleaves DNA at a specific sequence of nucleotides? The function of these enzymes, discovered in the 1960s and 1970s, is to destroy bacteriophages or other viruses that invade bacteria. Bacteria developed these enzymes to cut DNA sequences occurring within invading viruses. Once cut, the virus becomes harmless. The bacteria protects its own DNA from cutting by a restriction enzyme by chemically modifying its own DNA after synthesis. Many of the nucleotides in the bacteria's genome are chemically methylated. Once methylated, they are protected from digestion by the restriction enzyme. Thus restriction enzymes are capable of recognizing "nonself" DNA from "self."

This system of restriction endonucleases that recognize foreign DNA fulfills the criteria for being a primitive immune system for bacteria. How an enzyme evolved to achieve this degree of specificity in destroying viral DNA while preserving bacterial DNA is unknown. Restriction enzymes cannot work, however, as an immune system in humans because of the much greater size of the human genome. With 6 billion base pairs in human DNA (as compared with 1 million in bacteria), it is impossible to find a specific combination of nucleotides that occurs in a virus but does *not* occur very frequently in the much, much larger human genome. *Eco*RI, for

Table 3.1. Examples of commonly used restriction enzymes.

Enzyme	Recognition sequence	# sites in lambda
*Bam*HI	G/GATC[1]C	5
*Bgl*II	A/GATCT	6
*Dde*I	C/TNAG	>50
*Eco*RI	G/AA[1]TTC	5
*Hpa*II	C/CGG	>50
*Hind*III	A[1]/AGCTT	6
*Pst*I	CTGCA/G	18
*Sau*I	CC/TNAGG	2

[1]Will only cleave methylated DNA at that site. N stands for any nucleotide A, T, C, or G.

example, cuts human DNA into tens of thousands of fragments that are typically from 1,000 to 20,000 base pairs in length. For *Eco*RI to work as an antiviral agent in humans, all of these thousands of restriction sites would have to be methylated.

Table 3.1 lists a few examples of some of the more commonly utilized restriction enzymes. The enzyme, its recognition sequence, and the number of cut sites in the DNA of the lambda bacteriophage are given as a reference. Note that the larger the number of nucleotides in the recognition sequence, the less frequently the enzyme cuts lambda DNA. We could have predicted this from simple probability theory. The probability that a particular five-base sequence (given an alphabet of four nucleotides) will occur in a random series is $(1/4)^5 = 1/1,024$. So for an enzyme such as DdeI, a cut might be expected to occur every thousand or so base pairs. For an enzyme such as SauI, which recognizes a 7-base-pair sequence, the probability of finding a site is $(1/4)^7 = 1/16,384$. SauI is a "rare" cutter, breaking DNA into only big pieces, whereas DdeI cuts DNA into many more small pieces. The hundreds of restriction enzymes provide a very flexible system for engineered cuts in DNA.

The use of restriction enzymes as a tool in recombinant DNA technology can best be demonstrated with an analogy. The paragraph below contains a copy of the Preamble to the Constitution of the United States in which there is an error. We will use the technique of restriction enzymes to find the error. I bet that you cannot find it by yourself. Imagine a restriction enzyme that recognizes the sequence "the." Find within the Preamble every occurrence of the sequence of letters "the" and make a mark through the "t." Count the number of letters and spaces between each mark and write down these numbers. These numbers represent the lengths between each restriction "cut."

We the People of the United States, in Order to form a more perfect Union, establish Justice, insure domestic Tranquility, provide for the common defense, promote our general Welfare, and secure the Blessings of Liberty to ourselves and our Posterity, do ordain and establish this Constitution for the United States of America.

A friend in Washington, DC, who has done the same exercise while looking at the original Preamble to the Constitution tells you that his restriction "cuts" produce fragments of 4, 14, 28, 28, 32, 103, and 118 characters long. This is different from the result you obtained. Can you determine where the mutation in your copy of has occurred?*

Southern Blot

The exercise we have carried out in analyzing the Preamble to the Constitution for errors is equivalent to the method called Southern blot analysis. This method detects variations in DNA sequences by the relatively simple technique of cutting DNA with a restriction enzyme and seeing what size fragments are produced. To use our library analogy, imagine we were looking for the single error in the Preamble to the Constitution, but we did not know where in the library that particular piece of text was located. The Southern blot method makes this task feasible. Imagine that we had cut every bit of text within the entire library at the word "the." We would now have hundreds of thousands of pieces of text. We need to find the variation in text fragment lengths that occur only within the Preamble. We can do this by "highlighting" that portion of the text by hybridizing it to a marker.

Let me explain the actual process of a Southern blot by using human DNA. The DNA is extracted from a tissue sample; white blood cells from the peripheral blood are an excellent source of DNA. The DNA is incubated for several hours with a restriction enzyme such as *Bgl*II (pronounced "bagel two"). The enzyme cuts the entire human genome into tens of thousands of fragments ranging from 100 to 20,000 base pairs long. We now sort these fragments according to size by agarose gel electrophoresis. The small fragments of DNA move more rapidly than larger fragments by migrating through the pores in the gel. Unless stained with a dye such as ethidium bromide and viewed under ultraviolet light, DNA within an agarose gel is invisible. Figure 3.1 shows an agarose gel viewed under these conditions. Lane 1 contains size markers consisting of DNA fragments of known length. Lanes 2 through 9 each show a smear of DNA fragments of varying size of restriction digests of DNA samples from several patients. Note the small rectangles at the top of each lane. These are the wells into which the DNA was placed prior to electrophoresis.

The next step in a Southern blot is to blot the DNA from the gel onto a

---

*The error is in the 27th word; "our general Welfare" should be corrected to "the general Welfare." This error could have a significant impact on the interpretation of the Constitution and is a subtle but dangerous mutation! Because of the missing "the" restriction site in the mutant copy, your fragment lengths should have been 4, 14, 28, (28 + 32 =) 60, 103, and 118 characters long.

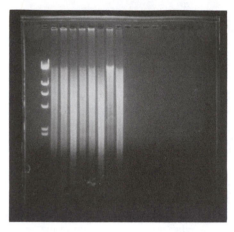

FIGURE 3.1. An agarose gel electrophoresis of DNA is photographed under UV illumination. Lane 1 shows several bands that serve as size markers. Lanes 2 to 9 show a smear of the thousands of DNA fragments that result from the digestion of human DNA with a restriction enzyme.

piece of electrostatically charged paper, which will bind the DNA. Filter paper made of nitrocellulose or certain types of nylon works well. The paper blot is much easier to handle than the fragile gel and will make the next step of hybridization much easier. The standard way to blot the gel is to place the nitrocellulose paper on top of the agar slab and then lay a pile of absorbent paper on top of the nitrocellulose. If the bottom of the agarose slab is sitting in buffer, the fluid will be drawn out vertically by diffusion. The DNA will be wicked out of the gel and stick to the charged paper. That's a Southern blot, named after Ed Southern. For laboratories that do many blots, it has become more convenient to use a vacuum blotting apparatus, which does the job in a few hours instead of overnight.

After the gel is blotted the next step is hybridization. A radioactively labeled probe that binds only to those DNA fragments that are of interest is incubated with the blot in a special buffer. The hybridization is usually carried out in a thin plastic bag that contains the paper blot and a few milliliters of radioactive probe in buffer. The probe is a cloned fragment of DNA that has a complementary sequence to the DNA fragments of interest. During an overnight incubation with gentle agitation, the probe seeks out and binds fragments to its target by base pairing. The specificity of binding of probe to target varies with what is called the stringency of the hybridization. Stringency can be adjusted by changing temperature, salt concentration, and other buffer molecules. Adjusting the stringency ensures the specificity for binding of the probe to target sequences. After incubation, the blot is washed and placed on x-ray film for a few hours and then developed. The resulting autoradiograph shows just the bands of interest.

Figure 3.2 is an example of an autoradiograph of a Southern blot. This blot is an analysis of the myeloperoxidase gene. Myeloperoxidase is an enzyme expressed in white blood cells of the granulocyte series. The amount of myeloperoxidase in granulocytic leukemias is abnormal, which is why I

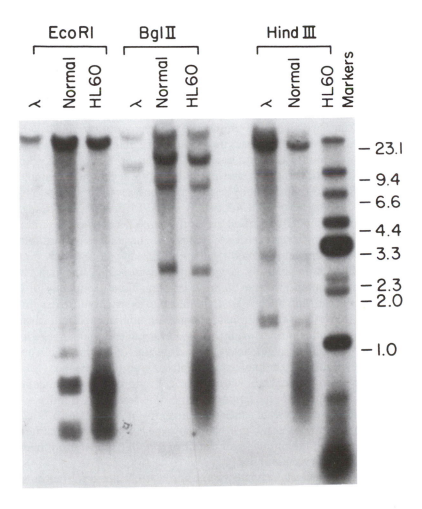

FIGURE 3.2. An autoradiograph of a Southern blot probed for the myeloperoxidase gene shows no difference between DNA from a normal subject and from leukemic cells.

was interested in probing this gene. The samples analyzed include DNA
from lambda bacteriophage, from a normal subject, and from a sample of
HL60 leukemic cells. Three different restriction enzymes were used: *Eco*RI,
*Bgl*II, and *Hind*III. Lane 10 contains molecular weight markers. The autoradiograph shows identical bands for the normal vs. the leukemic sample
for each of these enzymes. In this analysis, I could find no difference in the
myeloperoxidase gene between a normal subject and a leukemia. Southern
blot analysis, as will be shown in subsequent chapters, has been applied to
many other diagnostic problems in which gene analysis would be useful,
including sickle cell disease, cystic fibrosis, and the Philadelphia chromosome translocation in chronic myeloid leukemia.

Northern and Western Blots

The Southern blot method detects alterations in DNA sequences by looking
for a change in the pattern of DNA fragments after digestion with a restriction enzyme. Among the tens of thousands of fragments produced when
DNA is digested, we are able to see the very few fragments that interest us
by hybridizing the blot with a radioactively labeled probe. The Northern
blot is analogous to the Southern blot method but is a technique for analyzing messenger RNA. Similarly, the Western blot analyzes protein. The naming of these blots, Northern for RNA and Western for protein, is a joke on
the initial naming of DNA blots after Ed Southern. Whereas Southern blots
are a means of studying the structure of genes, Northern and Western blots
are a means of studying gene expression.

Messenger RNA is intrinsically a very unstable molecule, and it is this
fact more than anything else that makes Northern blots technically difficult. It is reasonable that mRNA is unstable; messages should exist only
long enough to have transmitted their information and then should disappear. Imagine the problem that would occur if every telephone message left
on your desk had failed to go away over a period of years. Each new
message would be lost in the background pile of the old messages. Most
mRNA is degraded by enzymes within the cell cytoplasm within a few
minutes after it comes off the polyribosome. The analysis of mRNA is
useful in that it shows us what genes are currently in operation in a cell.
Thinking of our library analogy, mRNA analysis is equivalent to telling us
what books have been taken out from the library today; it is not an inven-
tory of the entire library's contents.

A Northern blot analysis begins by isolating RNA. Cells are lysed in the
presence of strong RNAase inhibitors to prevent RNAase from destroying
the mRNA. Isolation of mRNA from this crude lysate is usually performed
by chromatographic separation. The mRNA is then electrophoresed
through an agarose gel. It is not necessary to digest the mRNA with a
restriction enzyme, because mRNA molecules are already small. After electrophoresis, blotting occurs as in the Southern technique followed by hy-

bridization with a labeled probe specific to the mRNA of interest. Technically, I have found the Northern blot to be difficult because of the extreme instability of mRNA.

The Western blot is an entirely analogous method for the analysis of proteins. Western blots are a bridge between recombinant DNA technology and more standard immunologic methods. Proteins are electrophoresed through a gel to separate molecules according to size, blotted onto a membrane, and then hybridized with an antibody against a specific protein of interest. In a Western blot the hybridization depends on antigen antibody binding, not binding of base pairs as is the case for DNA and RNA analysis. Figure 3.3 shows a Western blot analysis for the presence of antibodies to the human immunodeficiency virus (HIV). The presence of these antibodies in the blood of a patient confirms infection with the virus that causes AIDS. (See also Chapter 5.)

Southern, Northern, and Western blots start with an electrophoretic separation of a macromolecule. The macromolecule is DNA for a Southern, RNA for a Northern, and protein for a Western blot. There is a shortcut that we can take if we are relatively sure of the specificity of a particular analysis. Rather than electrophorese the macromolecule, we can just place a drop of the cell lysate onto a filter paper and carry out a hybridization directly on the spot. Such analyses are called "dot blots" or "slot blots," depending on the shape of the spot. Figure 3.4 shows a slot blot that measures the amount of c-*myc* DNA in HL-60 leukemic cells vs. the amount in normal lymphocytes. Since we were very sure of the specificity of our c-*myc* probe (that is, the probe was not binding to something else that looked like DNA from the c-*myc* gene, but wasn't), we took a shortcut in this experiment. The slot blot does not require the time-consuming steps of digestion with a restriction enzyme and electrophoresis. The slot blot in Figure 3.4 shows two parallel dilutions of DNA — one from HL60 leukemia cells and the other from normal lymphocytes. Each slot, as we progress down the ladder, contains one-half as much DNA as the slot above it. The blot has been probed for the c-*myc* oncogene. We can see that the signal in HL60 cells takes five more dilution steps to extinguish than for the signal in normal cells. These results demonstrate that HL60 leukemic cells have 32 ($= 2^5$) copies of the c-*myc* gene per cell, while normal lymphocytes have only one copy per cell. This is an example of gene amplification, which occurs in some tumors. (See Chapter 8.)

In Situ Hybridization

The hybridization of a labeled nucleic acid probe directly to cells or tissues is called in situ by hybridization. In this technique the target is the DNA or RNA in tissue sections or cell smears. For comparison, recall that in the Southern blot method DNA is extracted by lysing cells and is the electrophoresed and blotted onto a membrane. Hybridization between target DNA

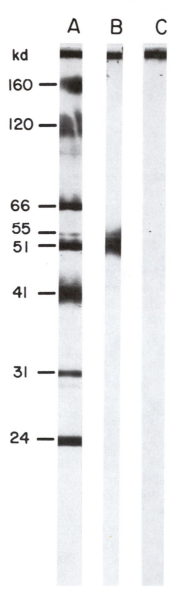

FIGURE 3.3. A Western blot analysis is shown of blood from three patients with possible HIV infection. Lane A shows the presence of multiple antibodies to HIV proteins indicating infection. Lane B shows the presence of a single antibody; this is a cross reaction and does not indicate HIV infection. Lane C shows no antibodies, also a negative result. (This Western blot is from the laboratory of Dr. James Folds, University of North Carolina, Chapel Hill.)

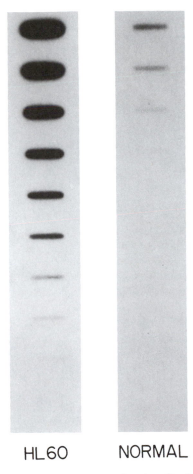

HL60   NORMAL

FIGURE 3.4. A slot blot of DNA from leukemic cells shows an increased number of copies of the c-*myc* gene in leukemic cells as compared to normal lymphocytes.

sequences and labeled probe occurs on the membrane. With in situ hybridization the target DNA sequences remain within the cells. Hybridization is carried out by incubating the probe with the tissue. After hybridization, the signal is read by looking directly at the tissues. One can tell which cells hybridize to the probe. Morphologic information is preserved along with the new information resulting from DNA analysis.

Figure 3.5 shows a photomicrograph of tissue culture cells that have been

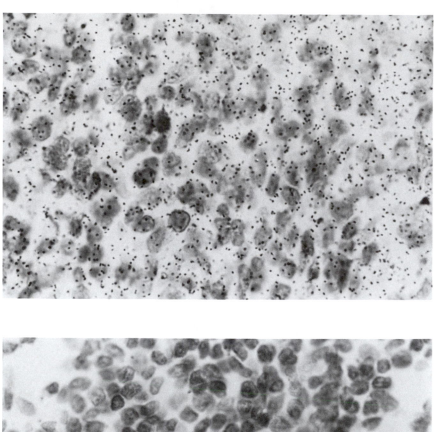

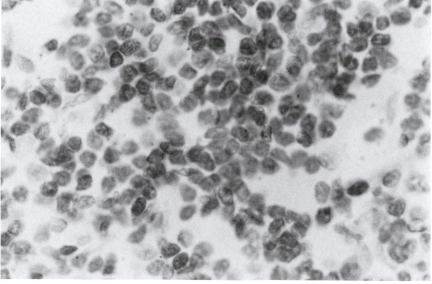

FIGURE 3.5. An in situ hybridization of tissue culture cells with a radioactive probe for a gene from the Epstein-Barr virus. The upper panel shows a large number of grains in the photographic emulsion that overlies the cells, indicating the binding of probe to target. The lower panel is a control in which a nonsense probe was used; no binding is seen. (This in situ hybridization is provided by Dr. Margaret Gulley, University of Texas, San Antonio.)

hybridized with a probe specific for the Epstein-Barr virus (EBV). In this experiment we are looking to see if one of the genes of EBV is being expressed, indicating an active vs. a latent viral infection. The top half of Figure 3.5 shows the probe bound to EBV mRNA in the cytoplasm of the cells. The bottom half of Figure 3.5 is a negative control performed with a "nonsense" probe. For the experiment shown in Figure 3.5, the probe has been labeled with a radioactive tag for increased sensitivity. Compare this in situ hybridization with Figure 8.9 (Chapter 8), in which we are looking for the presence of human papilloma virus (HPV). In this application the probe is tagged with a cytochemical marker for convenience, since the increased sensitivity of a radioactive probe is not necessary.

Labeling Nucleic Acid Probes

We have discussed hybridization for Southern and Northern blots, as well as for in situ methods, assuming the availability of labeled probes. The procedure for labeling a DNA probe with a radioactive tag is relatively straightforward. Radioactive probes are easy to detect by autoradiography and have a very high signal-to-noise ratio, giving the method great sensitivity. To produce a labeled probe, the probe DNA is incubated in a mixture of radioactive nucleotides. In a technique called nick translation, a DNA polymerase (whose function is to repair single strand breaks in the DNA molecule) is added to the mixture. The polymerase, along with tiny amounts of nuclease, spontaneously makes nicks and then fixes the nicks by using radioactive nucleotides to fill the holes. This makes the probe molecule radioactive. An alternate method for producing labeled probes is called random priming. In this procedure new pieces of DNA are synthesized by using the probe molecule as template. Again, the synthesis is done from a pool of radioactive nucleotide precursors. Both of these methods produce a radioactively labeled probe with high specific activity that gives a strong signal on autoradiographs. However, radioactive probes are stable for only 1 to 2 weeks. After that time, radioactive decay chemically breaks up the probe and it becomes less and less sensitive for hybridization.

Because of the instability of radioactive probes and some technical and safety issues in working with radioactivity, methods for labeling pieces of DNA by nonradioactive or "cold" labels have been devised. Instead of using a radioactively labeled nucleotide, you can use a nucleotide that is labeled chemically. Cold-labeled probes, after they are hybridized to target DNA, must be detected by cytochemical staining or other methods rather than autoradiography. Biotin, alkaline phosphatase, and chemiluminescent molecules are used to produce nonradioactive probes. The major advantage of cold probes is that they are stable and can be stored and used safely over a period of years. This is a particular advantage to commercial manufacturers wishing to include a labeled probe as part of a diagnostic kit. Each method for labeling DNA probes has slight advantages and disadvantages, and the variety of methods gives much flexibility in designing hybridization technologies.

Polymerase Chain Reaction

The polymerase chain reaction (PCR) is perhaps the single most important technique used in recombinant DNA analysis. The story of the discovery of PCR in 1983 by Kary Mullis, for which he received the 1994 Nobel prize in Medicine, is a fascinating one (see Mullis, 1990). Mullis was driving from his San Francisco lab to a weekend vacation along the Northern California coast. During the late-night drive, while his girlfriend slept, Mullis pondered a way to make repetitive copies of a single piece of DNA. By the time he had arrived at the cabin, he had thought out a chain reaction technique. Within months, he had a working method that is now standard around the world. These were the heady early days of recombinant DNA technology. Mullis revolutionized molecular exploration with one simple idea.

PCR works by selecting a fragment of DNA, typically 100 to several thousand base pairs long, and amplifying this fragment by repetitive cycles of DNA synthesis. Thus a particular DNA sequence of interest among the background of the entire human genome (6 billion base pairs long) can be amplified so that the small fragment becomes the majority of the DNA in the sample.

The method by which PCR works is as follows: Two small pieces of DNA, typically 20 base pairs long, called oligonucleotide primers, are synthesized. These primers are complementary images to each end of the DNA sequence of interest.

The primers plus an excess of free single nucleotides are added to the target DNA sample along with a heat-stable enzyme that promotes DNA synthesis called Taq polymerase.

The DNA mixture is now cycled through three different temperatures, which results in phases of the reaction: melting at 95°C, hybridization at 55°C, and synthesis at 75°C. Figure 3.6 demonstrates these three phases.

At 95°C, double-stranded DNA melts into two single strands. The thermal energy at 95°C is sufficient to overcome the hydrogen bonding between the base pairs of the two DNA strands.

After melting the reaction is cooled to 55°C, at which temperature the primers bind or hybridize to their complementary sites on the single-stranded DNA molecules. The temperature, pH, and salt content of the reaction favor binding of the primers, but do not encourage the entire single-stranded DNA molecule to reform double strands. In Figure 3.6, the two primers P1 and P2 are indicated along with the complementary regions on each DNA single strand cP1 and cP2. Note that at this point, all the DNA in the mixture is single stranded except for the two regions defined by the primers P1 and P2.

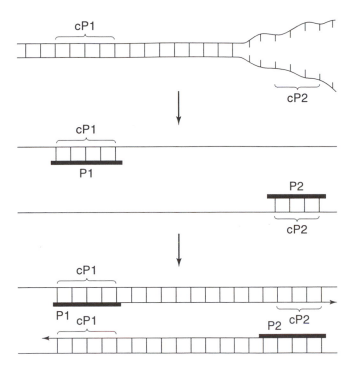

FIGURE 3.6. The polymerase chain reaction consists of three phases—melting, hybridization, and synthesis—controlled by thermal cycling of the reaction mixture.

The temperature is now raised to 75°C and new DNA synthesis begins. The Taq polymerase will promote DNA synthesis only beginning where the DNA molecule is double stranded. The synthesis proceeds from the double-stranded priming sites P1 and P2, copying the single strands in the 5' to 3' direction along the DNA backbone. The reaction typically runs at about 20 nucleotides per second. Within several minutes a new piece of DNA about 1,000 base pairs long has been copied from each strand. This completes the third phase of the first cycle of the polymerase chain reaction. Two new copies of the section of DNA of interest, between P1 and P2, have been created.

The second cycle begins by once again reheating the mixture to 95°C, which causes all the DNA to melt back to single strands. The temperature is lowered to 55°C and the primers hybridize to their targets cP1 and cP2. However, there are now four targets present, the original two plus the two new targets on the DNA synthesized in the first cycle. Figure 3.7 demonstrates how from this point on each cycle of the reaction doubles the number of DNA copies of the segment between cP1 and cP2. These cycles are repeated over and over, typically 30 times, in a device called a thermal cycler. At the end of 30 cycles of amplification there are typically 1 million

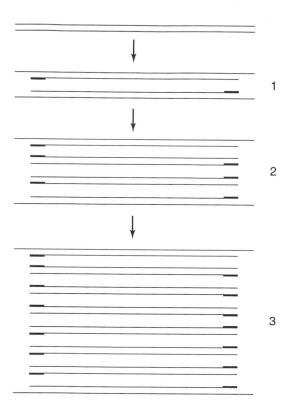

FIGURE 3.7. The polymerase chain reaction doubles the number of DNA copies of a segment of interest in each cycle. A typical 30-cycle reaction can amplify the target segment typically a million times.

copies of the DNA segment of interest for each original template molecule in the starting sample.

The Taq polymerase is a key ingredient of this PCR mixture. Taq (pronounced "tack") was isolated from the microorganism *Thermus aquaticus*, which was found growing in the hot springs of Yellowstone National Park's geyser basin. These organisms thrive at 75°C and have enzymes with maximum activity at this high temperature. Furthermore, Taq resists denaturation at 95°C, the necessary melting temperature of the reaction. When Mullis first designed the PCR method, he did not use Taq. Instead a conventional source of DNA polymerase was used. This enzyme required lower temperatures for synthesis and was destroyed in each cycle when the reac-
licenses it to other uses such as hospital laboratories as well as supply

tion was heated to 95°C. The first PCR methods required someone to stand over the mixture and pipette in more polymerase after each melting or about every twenty minutes for 30 cycles! The clever use of a high-temperature Taq polymerase means the reaction can cycle faster and no additional enzyme needs to be added. In modern PCR protocols, the mixture is prepared in a small plastic tube and placed in the thermal cycler for about 4 hours during which the 30 cycles of temperature change occur automatically under computer control. A thermal cycle is about the size of a breadbox and costs between $2,000 and $14,000 depending on options.

Another feature that makes PCR so easy is the availability of primers. Short pieces of DNA are very easy to synthesize. Automated machines produce primer overnight from a typed-in sequence of the 20 or so desired nucleotides.

The PCR method only produces a successful amplification when the two regions cP1 and cP2 are within 1,000 to 2,000 base pairs of each other. The DNA synthesis in the third phase must be sufficient to span the gap between cP1 and cP2. If the primers are very far apart such that the newly synthesized strands do not span the gap, the number of copies in each cycle will increase arithmetically rather than geometrically. Thus, at the end of 30 cycles there will be only a trivial increase of 60 additional copies, instead of 1 million or more. The choice of primers controls the amplification of the segment of interest.

At the end of the polymerase chain reaction, an agarose gel electrophoresis is performed. The amplified product should appear as a band of a single molecular weight in the gel. Figure 3.8 is an example of the results of a PCR analysis. At either end are lanes containing molecular weight markers. The bright bands in many of the lanes are amplified segments of DNA. This gel electrophoresis can be skipped in some applications. When a successful amplification has occurred, the PCR mixture changes from clear to cloudy, indicating the presence of a large amount of larger-sized DNA pieces.

The PCR method is capable of identifying a specific gene segment (such as the presence of HIV within a blood sample) within a single day. There are, however, some artifacts that can distort the analysis, and care must be paid to technique. The most important known artifact is that of contamination by minute quantities of DNA carried over from previous samples. If, for example, in looking for the presence of the hepatitis virus genome in a human blood sample, a very small amount of DNA (down to one part in 10^6) were carried over from a previous positive patient sample, then this DNA would be sufficient to prime the PCR reaction. The carryover of a contaminant would result in a false-positive result. The extreme sensitivity of PCR is both its strength and at times its weakness.

PCR was invented by Kary Mullis while he was at the CETUS Corporation (Emeryville, CA). Licensing for diagnostic medical applications has been granted by CETUS to Hoffmann-LaRoche (Nutley NJ), which in turn

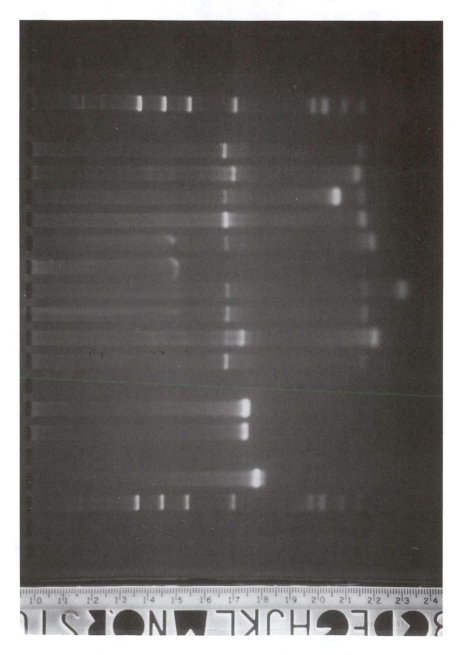

FIGURE 3.8. An agarose gel electrophoresis of the products of a PCR. Lanes 1 and 18 contain molecular weight markers; lanes 2 to 17 show bands that represent specific amplification of DNA fragments. The bright band in the middle of most lanes is due to the amplification of a control gene. The fainter bands near the bottom of the gel in some lanes indicate amplification of the target gene under investigation. (This photograph is provided by Dr. Steve Schichmann, University of North Carolina, Chapel Hill.)

diagnostic kits and services. There are other amplification technologies, such as beta Q replicase, transcription-mediated amplification (TMA), and strand displacement amplification (SDA), which may become competitors for PCR. However, PCR has spread throughout the world in a very short time and is the current mainstay of recombinant DNA technology.

Cloning

One of the most important methods of recombinant DNA technology is gene cloning. Cloning means isolating a gene of particular interest and making copies of it. Once isolated and cloned, a gene can be further modified by genetic engineering techniques to take on new properties. Figures 3.9 and 3.10 schematically demonstrate the steps in cloning. One begins with a piece of the gene or a DNA sequence of interest. The DNA segment is cut out with restriction enzymes and inserted into a plasmid, as is shown in Figure 3.9. Then the plasmid is grown in the appropriate bacteria and produces millions of additional exact copies, as shown in Figure 3.10. Finally, the gene sequence of interest is removed from the plasmid and is available for further use as a DNA probe or for genetic engineering. Figure 3.10 shows the steps involved in producing large quantities of the plasmid DNA and its cloned insert for some industrial or pharmaceutical applica-

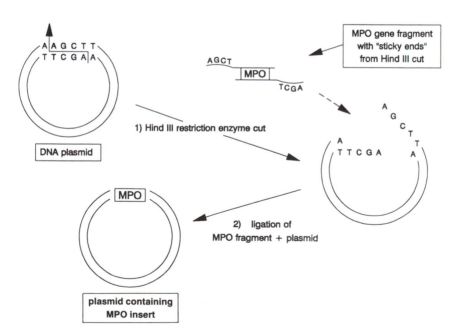

FIGURE 3.9. The insertion of a DNA segment into a plasmid as demonstrated here is a key part of gene cloning.

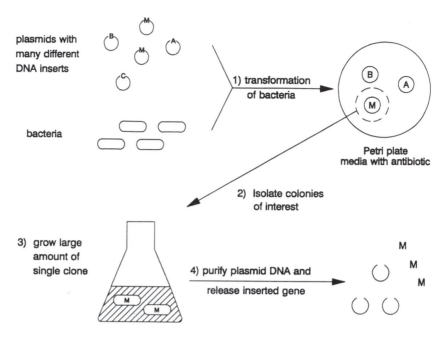

FIGURE 3.10. The process of cloning a gene involves a numer of steps schematically demonstrated here that involve both technology and some clever thinking on
the part of the investigator.

tions. The amount of bacteria and plasmids cultured may be in hundred-
liter quantities, producing grams of DNA.

I will consider each of the steps in cloning in more detail. Certainly one
of the most difficult and artful steps is the first one, finding the gene of
interest. Let me give a specific example, which I will use to demonstrate the
entire cloning process. I wish to clone the gene for myeloperoxidase. This is
an enzyme present in neutrophils that aids in killing bacteria. Cloning the
gene will give me a probe so that I can study the regulation of myeloid
maturation in bone marrow cells. Cloning the gene will also give me the
means of producing the enzyme by recombinant DNA technology for possible use as an antibacterial agent. With good reasons for wanting to clone
this gene I begin my work. If all goes well, in a good research laboratory
the process takes a few months.

To begin, I have to find the gene sequence for myeloperoxidase, located
somewhere within the entire human genome. This is equivalent to trying to
locate a specific paragraph in a book when you have an entire library to
search through and no card catalog, index, or other guide. This is where the
artful tricks in gene cloning begin. I know that neutrophils contain a lot of
myeloperoxidase. This implies that the granulocyte precursor bone marrow

cells are synthesizing this gene product at a high rate. I limit my search by creating what is called a "cDNA library" made from cells that are expressing myeloperoxidase.

I know there is a cell line of cultured promyelocytes called HL60 cells. I acquire these cells and grow them in tissue culture. I verify by conventional enzymatic staining methods that these cells are making large amounts of myeloperoxidase. To construct my library, I isolate the mRNA from these cells. I keep only the mRNA, discarding the DNA and proteins. The mRNA consists of copies of all the genes that are turned on in promyelocytes. These cells are making a large number of proteins, but we know that one of them is myeloperoxidase. The number of different types of mRNA may be in the hundreds, representing messages from all the genes that are currently active. However, this is certainly less than the 300,000 genes present in the human genome. Thus, by using mRNA I have limited my search from several hundred thousand genes to several hundred. With luck, since myeloperoxidase is being synthesized in large amounts, I can hope that a fairly high percentage of the pieces of mRNA that I have isolated are messages for myeloperoxidase, making my search easier.

I now take the mRNA, and using an enzyme called reverse transcriptase, synthesize a DNA copy of each mRNA molecule. This DNA is called complementary DNA or cDNA. It does not copy the genes as they would appear in the original DNA from the cell. cDNA is a direct copy of mRNA and as such contains only information present in the exons of the gene. All the introns are removed. A cDNA copy of a gene is a more compact copy than the original. cDNA contains information for the protein but does not contain the leading promoter sequences or any of the spacing introns. I work with cDNA because it is much more stable than mRNA and because it can easily be inserted in plasmids.

Actually, everything that I have done so far in cloning the myeloperoxidase gene is so standard that I really would not have had to do it. cDNA libraries are a very useful resource to a molecular biologist and are available from colleagues or commercial laboratories. To save time and considerable money, I obtained my cDNA genomic library of HL60 promyelocytes from a commercial supplier for $450. Looking through the catalog, I had a choice of many libraries including human fetal brain, human adult pancreas, mouse T lymphocyes, insect salivary gland, and many others. I chose HL60 promyelocytes because these cells are synthesizing myeloperoxidase and therefore must be making a lot of mRNA from this gene. If I had been looking for a gene that was associated with, say, brain development, I would have purchased the fetal brain library.

The fragments of cDNA are carried as inserts within plasmids. Plasmids are like viruses. They live parasitically in bacteria. Recombinant DNA technology has greatly modified plasmids to make them useful for the cloning process. Plasmids serve as carriers for pieces of DNA. An analogy is to think of a plasmid as a computer floppy disk. A special piece of DNA is carried by a plasmid just like a computer program of interest is written on a floppy disk. Once the DNA is in the plasmid, it is very simple to make large number of copies. Many engineered plasmids are available for cloning; a few are very commonly used, like pBR322 and M13 (see Table 2.1).

To insert my cDNA into plasmids, I cut it with a restriction enzyme. This produces fragments with a specific "sticky" cut end. The cut ends are called sticky because under the right conditions they will rapidly reseal with each other. In a separate test tube, I take a plasmid and cut it with the same restriction enzyme, producing sticky ends identical to those on my cDNA fragments. I now mix my cut cDNA and my cut plasmid together, as shown in Figure 3.9, and incubate them in conditions that favor the reannealing of cut DNA. This is accomplished by adding an enzyme called a ligase. The plasmids reanneal, but sometimes a piece of cDNA gets inserted into the plasmid. The cDNA fragments have the same cut ends and fit into the plasmid easily. A plasmid with a DNA insert is frequently called a vector.

The engineered plasmid that we have used carries a gene for antibiotic resistance. After I fuse together my plasmid vectors with host bacteria, I will grow the bacteria in medium containing an antibiotic. Only bacteria that have a plasmid inside them will have the ability to grow in the presence of the antibiotic. This eliminates any bacteria that fail to fuse with a plasmid from my culture. The bacteria are grown on petri dishes inoculated with a dilute solution, so that only a few hundred colonies will grow. From these few hundred colonies, I must find which few contain plasmids with a piece of the MPO gene. I have shown this schematically in Figure 3.10.

This step of finding the one bacterial colony with a piece of the MPO gene is potentially difficult. I must search each colony, one by one, for pieces of cDNA that are part of the myeloperoxidase gene. The way I find the myeloperoxidase gene is by knowing a little bit of the structure of the protein. A portion of the amino structure of the protein is known. I choose six amino acids in the middle of the protein, and using the genetic code, I calculate which six DNA codons would be present in the gene to specify these amino acids. I then synthesize a short piece of DNA called an oligonucleotide probe that represents these six codons. My oligonucleotide probe is 18 bases long (three bases for each of the six codons). My probe is a small fragment of the genetic message that must be part of the myeloperoxidase gene.

I will use this small oligonucleotide probe to find much bigger fragments of the myeloperoxidase gene in my cDNA-containing plasmids. I do this by blotting the surface of the petri plate onto a circle of nitrocellulose paper. DNA is firmly bound to the electrostatically charged nitrocellulose paper. Under the appropriate conditions similar to a Southern blot, I incubate the circular blot of the petri plate with my oligonucleotide probe. The colonies that contain plasmids having a portion of the myeloperoxidase gene inserted in them hybridize with my probe. Of the hundreds of colonies on the plate, perhaps two or three hybridize with the probe. I go back and pick these colonies off the plate and put them in small test tubes.

After I have picked ten or so colonies, I go to the next step. I release the inserts from the plasmid and measure the size of the cDNA fragments by electrophoresis. Each of these pieces has the 18 nucleotides that were present in my probe, but they also contain a larger portion of the entire myeloperoxidase gene. I take the largest of my cloned fragments and in turn use it as a probe to screen more bacterial plates. This allows me to isolate other fragments of the cCNA that contain a portion of the myeloperoxidase gene further away from my initial probe. Eventually, I have tens of cloned fragments of overlapping pieces of the myeloperoxidase gene. By a tedious process that I will not detail here, I sort these fragments out until I have covered the entire length of the myeloperoxidase gene.

I know when I have the entire myeloperoxidase gene because of my knowledge of the anatomy of a gene. I must look for a stop codon at the end of the gene and also find start sequences near its beginning. When I have a single fragment that contains the whole sequence from start to finish, I have successfully acquired a cDNA copy of the myeloperoxidase gene.

My cDNA clone of the myeloperoxidase gene is quite a useful product. I can make many copies for distribution to other research labs. I can insert this cDNA piece into yeast. If I place a strong promoter ahead of the gene, the yeasts, as they grow, will synthesize large amounts of the myeloperoxidase gene product. Although myeloperoxidase is not part of the yeasts' natural genome, these engineered yeasts will now synthesize this protein. I can produce pharmaceutical quantities of myeloperoxidase for study as an antibacterial agent.

Now that I have a cDNA copy of the myeloperoxidase gene I can find its location within the genome. I can quickly discover that the myeloperoxidase gene is located on chromosome 15 at a location called 15q3.1. This will be of great help to the human genome project as a signpost to watch for at that site. I would certainly sequence my cDNA and submit the complete genetic code to the Human Genome Bank.

It seems that nearly every day you read in the newspaper or in a medical journal that a laboratory has discovered a new gene. Understanding the process of cloning, you can read the article with more sophistication and ask (1) Did they clone the gene and have a complete cDNA fragment? (2) Have they localized only a small portion of the gene and are they now in the process of cloning it? (3) Have the localized the gene to a specific chromosome site? (4) Have they sequenced the gene and do they know its detailed structure? (5) Do they know what protein the gene is making and understand its function in the cell?

When you read about the discovery of a gene, it usually represents only one of these steps. That means much more work is to be done. Sometimes a gene is discovered inadvertently but its function is completely unknown. Other times when a gene is discovered, what is really meant is that a small portion of it has been found and a great deal of work remains to clone the entire sequence and have it fully characterized.

Transgenic Animals

The creation of transgenic animals is one of the most dramatic advances derived from recombinant DNA technology. A transgenic animal results from the insertion of a foreign gene into an embryo. The foreign gene becomes a permanent part of the host animal's genetic material. As the embryo develops, the foreign gene may be present in many cells of the body, frequently including the germ cells of the ovary or testis. If the transgenic animal is fertile, the inserted foreign gene (transgene) will be inherited by future progeny. Thus, a transgenic animal, once created, can persist into future generations. Transgenic animals are different from animals in which foreign cells or foreign organs have been engrafted. The progeny of engrafted animals do not inherit the experimental change. The progeny of transgenic animals do.

The techniques for creating a transgenic animal involve (1) choosing a foreign gene, (2) placing the foreign gene in a suitable form called a "construct," which guides the insertion of the foreign gene into the animal genome and encourages its expression, and (3) injecting the construct into a single fertilized egg or at the very early embryo stage of the host animal. Since the efficiency of the insertion of a foreign gene is low, there must be a selection process to identify when the gene has successfully incorporated into the developing animal's DNA.

Much genetic engineering goes into the choice of a foreign gene and building a construct. The construct must have promoters to turn on foreign gene expression at its new site within the host animal genome. By choosing a particular promoter and splicing it in front of the foreign gene, we can encourage expression of our transgene within a specific tissue. If, for example, a promoter from the immunoglobulin gene is used in the construct, then the foreign gene will be expressed preferentially in the B lymphocytes of the transgenic animal.

The process of creating transgenic animals has led to an improved understanding of the elements that control gene expression. A transgenic animal may express the foreign gene in only some of its tissues and only at limited times during development. The transgene will be expressed more consistently if a powerful promoter is included in the construct inserted into the embryo. The inserted gene in a transgenic animal may incorporate at a random site in the genome, or it may be carried within the cell as a freestanding bit of DNA. Recently, techniques have been developed to target the transgene to a specific site, such as adjacent to a defective gene. The process of creating transgenic animals has shown that there are no barriers to mixing DNA from one species into another. Human genes are routinely inserted into mice.

One of the most important applications of transgenic animals is the development of new animal models of human disease. Transgenic animals can serve as models for a number of malignant tumors. Inserting the c-*myc* oncogene, which regulates cell growth, into a mouse creates a transgenic strain with a high rate of spontaneous tumors. The type of tumor depends on the promoter placed in front of the c-*myc* gene in the construct. The mammary tumor virus (MTV) promoter increases the incidence of breast adenocarcinomas. The immunoglobulin heavy-chain enhancer (IgH), when inserted along with c-*myc*, results in a strain of mice with a high incidence of lymphoblastic lymphomas.

Creating transgenic animal models of inherited human diseases has been more difficult, but successful nonetheless. Placing a defective hypoxanthine guanine phosphoribosyl-transferase (HPRT) gene in a transgenic mouse did result in an animal with a homozygous HPRT deficiency mimicking the biochemistry of the human neurologic disorder, Lesch-Nyhan syndrome. Inserting a defective collagen gene into a transgenic mouse produced a perinatal lethal condition similar to osteogenesis imperfecta in humans. Efforts are under way to model many other human diseases in transgenic mice including neoplasia, immune defects, hypertension, neurologic disorders, and biochemical defects.

Transgenic animals also serve as models for gene therapy. Will a good copy of a gene correct an inherited biochemical defect when transfected into an organism? The human beta-hemoglobin-chain gene has been inserted into a mouse resulting in production of human beta chains. If this process could be used to insert a hemoglobin gene into bone-marrow stem cells taken from a patient with thalassemia or sickle cell anemia, then reinfusion of the "transgenic" bone marrow would correct the hemoglobinopathy.

Although mice have been the most frequent hosts for transgenic modification, other domestic animals have also been used. One surprising idea has been to create transgenic cows that secrete important pharmaceutical substances in their milk. The principle has been demonstrated by the development of a transgenic mouse strain that secretes tissue phasminogen activating factor (TPA) in its milk. Transgenic plants may also be a very useful source of new drugs. Work on modifying the tobacco plant and other species has shown that the plant can produce new proteins. Future crops of transgenic plants may be grown as a source of new pharmaceuticals. Transgenic animals provide models for many human diseases and specific pathologic processes at the level of the gene. Transgenic animals are tools for studying the basic biology of gene expression and modification of the genome via gene therapy for the purpose of curing disease.

Conclusions

The basic science of recombinant DNA technology is a fascinating story in its own right. Technological developments continue at a frenetic pace. One astounding example is the marriage of DNA hybridization with semiconductor technology in a process developed by AFFYMAX (Mountain View, CA). The company claims a patented process for the deposition of a matrix of multiple nucleic acid fragments on a semiconductor substrate. This leads to the fabrication of a microchip that when immersed in a solution can electronically signal the presence of specific DNA fragments. AFFYMAX was recently purchased by GLAXO. Our interest is in the application of recombinant DNA to medicine. Part II discusses numerous examples of molecular medicine. You will find, I believe, that some understanding of the basic science of molecular biology is useful in examining these medical applications. However, it is by no means necessary to be a molecular biologist to employ applications of molecular medicine in clinical practice. Like computer technology, recombinant DNA technology can be used by nonexperts. During the transition years, as molecular medicine is finding its rightful place within the science and art of medicine, some scientific knowledge will be useful to critique each new discovery. However, I find that it is much easier to teach molecular biology to physicians in order that they may employ it in the care of their patients than it is to explain medicine to molecular biologists who wish to know how their science can be applied to human health.

Bibliography

Erickson RP (1988) Minireview: Creating animal models of genetic disease. Am J Hum Gen 43:582–586.

Erlich HA (ed) (1989) *PCR Technology*. Stockton Press, NY, pp. 1–244.

Erlich HA, Gelfand D, Sninsky JJ (1991) Recent advances in the polymerase chain reaction. Science 252:1643–1651.

Jaenisch R (1988) Transgenic animals. Science 240:1468–1474.

Lathe R, Mullins JJ (1993) Transgenic animals as models for human disease. Transgenic Res 2:286–299.

Merlino GT (1991) Transgenic animals in biomedical research. FASEB J 5:2996–3001.

Mullis KB (1990) The unusual origin of the polymerase chain reaction. Scientific Am April 1990, pp. 56–65.

Nabel EG, Nabel GJ (1993) Direct gene transfer: basic studies and human therapies. Thromb Haemost 70:202–203.

Piper MA, Unger ER (1989) *Nucleic Acid Probes: A Primer for Pathologists*. ASCP Press, Chicago.

Rosenthal N (1994) Molecular medicine. N Engl J Med 331:315–317, 599–600.

CHAPTER 4

Cytometry — Cell Analysis

Introduction

Recombinant DNA technology, as reviewed in the last chapter, is designed to detect alterations in the genome down to the level of a single nucleotide alteration. Sometimes, however, we want a broader analysis of the DNA complement of a population of cells. The technology of automated single-cell analysis, flow or image cytometry, fills this need and bridges the gap between molecular analysis and examination of the whole cell. Cytometry is an older technology than recombinant DNA, and has well-established applications in the areas of cancer diagnosis and immunology. Cytometry labs are becoming common in larger medical facilities. This chapter examines the tools of cytometry including the flow and image systems currently available. The biology of the cell division cycle, which is the basis of DNA ploidy and S phase measurements in tumors, is presented first followed by clinical examples in cancer diagnosis. The application of cytometry to immunology is presented briefly and expanded upon in the section on AIDS in Chapter 5.

Let's begin, however, by realizing the difference in scale between detecting flaws in single nucleotides within the genome and examining a whole cell in which a large amount of damage has occurred, resulting in abnormal or extra chromosomes. To recall our analogy comparing the human genome to a university library (Chapter 1), a genetic mutation is the equivalent of a misspelled word. A cell with an abnormal number of chromosomes is the equivalent of a whole floor of the library being trashed!

An aneuploid cell is a cell that contains a quantitatively abnormal total DNA content as opposed to a euploid or normal DNA content. A euploid normal cell is also called diploid, meaning that the cell has the normal double copy of all the chromosomes (including XX or XY for female or male). The DNA content of a human diploid cell is 7.2 picograms of DNA per cell. The measurement of cellular DNA content by cytometry is called "ploidy" and is reported as a DNA index, where diploid DNA content is defined as a DNA index of 1.00. Table 4.1 defines some of the common terms used in ploidy determinations by cytometry. *Aneuploid* is a general term that refers to any cell with abnormal DNA content, which includes hypo- or hyperdiploid, tetraploid, and multiploid.

Another way to examine the DNA content of a cell is to visualize and count its chromosomes, called karyotypic analysis. A karyotype of a cell has the advantage of demonstrating specific chromosome abnormalities in aneuploid cells. Many chromosome abnormalities are associated with a specific medical syndrome. Figure 4.1 is a karyotype of a single cell that has an extra (third) copy of chromosome 21. Trisomy 21 is diagnostic of Down syndrome. Although this cell is hyperdiploid, flow cytometry would be unable to detect this small additional amount of DNA represented by an extra chromosome 21. An extra copy of chromosome 21 is only about a 1% increase in the total DNA content of the cell, a DNA index of 1.01. This is below the threshold of sensitivity for flow cytometry as we will discuss later. However, from the point of view of recombinant DNA analysis, an extra chromosome 21 is an additional 60 million base pairs! This is too big an error to be easily measured by Southern blot or PCR techniques.

Cytometry, chromosome analysis, and recombinant DNA technology detect vastly different size defects in the human genome. These methods also differ greatly in the number of cells they examine. Table 4.2 compares these technologies and demonstrates how together they bridge the gap between the molecular and the cellular.

Cell Division Cycle

In a growing cell all of the biochemical components, such as the protein and the DNA, must be synthesized to double the original amount before the cell can divide. Each daughter cell receives half, returning the amount of protein, DNA, and cell volume back to the normal "resting" level. Figure 4.2 depicts how the increases in cell volume and DNA occur over time during a typical cell division cycle, which in human cells can take from 10 to 24 hours. Note that while the cell volume increases smoothly and continuously over the entire cycle, DNA is only synthesized during a mid-

Table 4.1. Terms used for DNA ploidy measurement.

Term	DNA index (DI)	Chromosome type
Diploid (euploid)	= 1.00	Normal chromosomes, 46
Aneuploid	≠ 1.0	
Tetraploid	= 2.0	Double set of chromosomes, 92
Hypodiploid	< 0.9	Chromosome loss
Hyperdiploid	> 1.1	Chromosome gain
Multiploid	≠ 1.0	Multiple populations with abnormal DNA

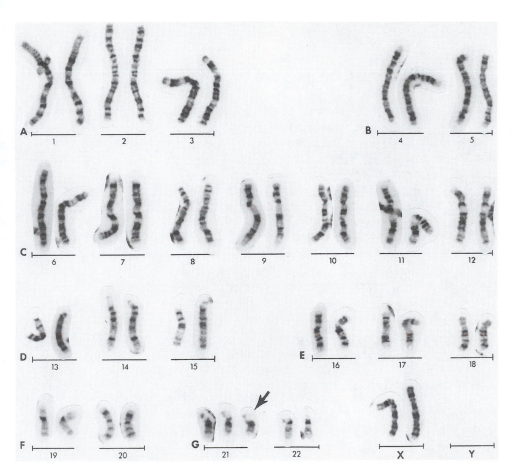

FIGURE 4.1. Chromosome karyotype of a single cell with an extra chromosome 21, characteristic of Down syndrome. (Photo courtesy of Drs. K. Rao and K. Kaiser-Rogers, Cytogenetics Lab, UNC, Chapel Hill, NC.)

Table 4.2. Molecular to cellular technologies.

Technique	Size of defect in nucleotides	No. of cells analyzed
Recombinant DNA	10^1-10^4	$>10^6$
Chromosome analysis	10^8	10^1
Cytometry, image	$>10^9$	10^2
Cytometry, flow	$>10^9$	10^4

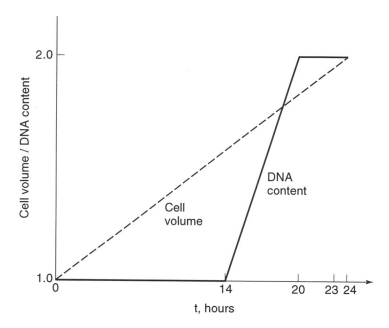

FIGURE 4.2. Cell division cycle. Cell volume increases smoothly throughout the cycle, while DNA is synthesized only in the mid-portion or S phase. The periods before and after the S phase are called G1 and G2. At M, the cell divides by mitosis into two daughter cells.

portion of the cycle, called the S phase. The periods prior to and after the S phase are called G1 and G2, the G standing for gap in DNA synthesis. After the G2 phase, the cell undergoes mitosis, M, which takes about 0.5 hours. Cells that are not dividing "rest" in G1. (Some researchers prefer to call the resting phase G0, and change the name to G1 as soon as the cell starts growth.)

Cytometry can easily measure both cell volume and DNA content. This is the method by which cell division is usually assessed. The instruments and methods of cytometry are presented in the next section of this chapter, but to complete our description of the cell division cycle we jump ahead to an initial look at some cytograms. A cytogram is the graphic output of a cytometer. One of the most useful cytograms is a frequency distribution of the DNA content. Two examples of DNA cytograms are shown in Figure 4.3. The DNA frequency distribution is the number of cells plotted on the Y axis vs. the cellular DNA content plotted on the X axis. Figure 4.3a is a DNA cytogram of a very slowly growing population of cells, such as peripheral blood lymphocytes. The cytogram has only one peak, since almost all

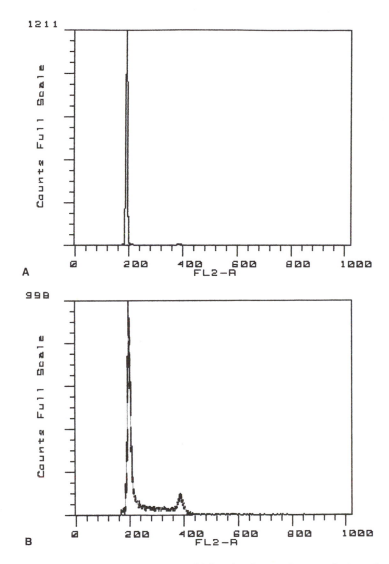

FIGURE 4.3. DNA cytogram of (a) a slowly growing population of peripheral blood lymphocytes, and (b) a rapidly growing population of lymphocytes in a malignant lymphoma.

the cells are in the G1 resting phase. Figure 4.3b is a DNA cytogram of a rapidly growing population of cells, a malignant lymphoma. This cytogram has three peaks representing the G1, S, and G2 + M regions, respectively. many of the cells, about 55%, are still in G1 since this is the longest phase of the cycle. About 37% are in the S phase and 8% in the G2 + M phase.

The G2 and M phases of the cycle are usually combined when reporting cytometry results because both G2 cells and mitotic cells have a DNA index of 2.0 and cannot be distinguished by this technique. Note that the G1 peak is quite sharp in contrast to the much broader S peak.

Cells in G1 all have DNA index of 1.0. Cells in S phase start with a DNA index of 1.0 and over 6 to 8 hours increase to 2.0. The mixture of cells with varying DNA amounts in the S phase makes the peak quite broad. There are mathematical models of the cell division that predict the distribution of DNA content. These mathematical models are used as part of the software of a cytometer instrument to calculate the cell cycle parameters from the measured DNA cytogram. This becomes particularly important, as we will see, in the analysis of tumors that frequently contain mixtures of normal and abnormal populations.

Flow Cytometry

A flow cytometer is the basic instrument of single-cell analysis. The technology is nearly 25 years old, although clinical applications have increased sharply only within the last 5 years. A flow cytometer analyzes a liquid suspension of cells by passing them single file through a laser beam. The scattered light and cell fluorescence are measured as the cell interrupts the laser beam, giving a measure of cell size, DNA content, or other cellular parameters. An alternate approach is image cytometry where the cell is viewed on a conventional glass slide using a microscope, digital camera, and a computer to measure individual cell characteristics. Image cytometry, which is a newer technology, is presented in the next section.

Figure 4.4 is a diagram of a flow cytometer. The basic components of a flow cytometer are (1) a fluidics system to aspirate and transport the suspension of cells; (2) an optical system consisting of a laser, lens, color filters, and photodetectors; and (3) a computer to collect the data in a digital mode and to calculate and display the results as a cytogram. The key principles that make flow cytometry possible are the physics of light scattering. The light that is scattered at low angles (around 5° off the optical axis) is closely related to the size of the cell. The higher angle scatter (around 10° to 15°) is a measure of the internal complexity of a cell and can help, for example, to distinguish a lymphocyte from a granulocyte. In addition to light scatter, we can stain the cell with a dye such as propidium iodide, which binds quantitatively to DNA. Propidium iodide is fluorescent, and by measuring the amount of fluorescence we can assess the DNA content. A flow cytometer typically measures these three parameters – low angle scatter (cell volume), high angle scatter (cell granularity), and fluorescence – simultaneously on each cell as it passes through the laser beam. The speed of the electronics is such that 2,000 cells can be analyzed per second.

The operator of the flow cytometer has interactive control over the analy-

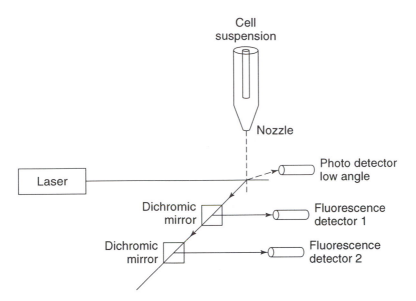

FIGURE 4.4. Schematic diagram of a flow cytometer.

sis. The tissue being analyzed in a clinical sample is frequently a mix of several cell types, only one of which may be relevant. For example, in a DNA analysis of a breast tumor, cells present in the sample would include breast carcinoma cells, nonmalignant breast epithelial cells, stromal cells, and blood cells. The measure of DNA ploidy and cell cycle parameters is of interest only for the breast carcinoma cells. The operator can focus on this cell type of interest by the process of "gating," which selects a subpopulation of cells passing through the cytometer based on cell size, DNA content, or other features such as fluorescence due to binding with an antibody to cytokeratin. By selecting a subpopulation based on one or more of these features, we can look at the cell type of interest.

Figure 4.5 is a DNA cytogram of an infiltrating carcinoma of the breast. This cytogram was produced by disaggregating a 0.5-cm cube of the breast tumor into a single-cell suspension, staining the cells with propidium iodide, and passing the cells into a FACSCAN (Becton-Dickinson, San Jose, CA) flow cytometer. The operator gated on large volume, cytokeratin-positive cells. The cytogram demonstrates a population of cells with normal DNA content (DNA index 1.0) and a second population with abnormal, aneuploid DNA content (DNA index 1.85). Using the computer programs that model cell cycle parameters, we can calculate the percent of cells in G1, S, and G2 for the abnormal population. The model estimates 6% of the tumor cells are in the S phase of the cycle. These parameters, aneuploid DNA

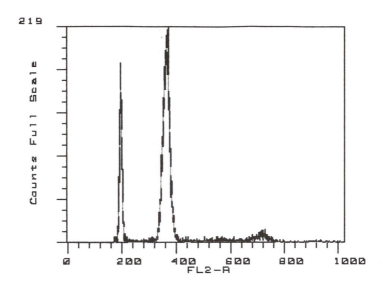

FIGURE 4.5. DNA cytogram of a breast carcinoma showing aneuploid DNA content in the G1 peak at channel 365. Some diploid cells with a normal G1 peak at channel 200 are also present, due to benign stromal cells and lymphocytes mixed in with the tumor.

content with normal S-phase fraction ($<7\%$), suggest that this patient's breast tumor will have intermediate biological aggressiveness. The presence of aneuploidy with an abnormally high S phase would indicate high biological aggressiveness. DNA ploidy and S-phase fraction, along with histologic grade and pathologic stage of the tumor, have prognostic importance and help guide therapeutic choices concerning surgery, radiation, and chemotherapy.

Flow cytometry also has major applications in immunology. Any flourescent-tagged antibody binding to a cell can be quickly quantified by flow cytometry. Thus, analysis of peripheral blood lymphocyte populations for the number of B, T, T-helper, T-suppressor, and NK-lymphocyte subclasses is routinely carried out by flow cytometry. The quantization of lymphocyte types is critical for AIDS therapy. More complex "immunophenotypes" encompassing a dozen or more surface markers can be carried out on lymph node tissue, and is useful for the classification of lymphomas.

Flow cytometry instruments are available from a number of companies including Becton-Dickinson (San Jose, CA), Coulter (Hialeah, FL), and Ortho Instruments (Raritan, NJ). The methods of flow cytometry and its quality assurance are becoming standardized (much more so than any other DNA analysis procedure). A major limitation of flow cytometry is one of sensitivity. Abnormal cell populations that are less than 5% different from normal cells will not be reliably detected. For example, a cell population with one extra chromosome, as in the case of trisomy 21 mentioned earlier, will not be recognized. A tumor sample with only a few malignant cells scattered within a reactive stroma would also not be seen. Image cytometry, described next, is a partial solution to this limitation.

Image Cytometry

Image cytometry has the same objective as flow cytometry—automated analysis of single cells—but uses a very different technology. An image cytometer is an automated microscope, and although the idea is not new, the recent availability of powerful, relatively inexpensive computers and digital video cameras has made the method feasible. Figure 4.6 is the digital image of a group of cells stained with a dye that binds to DNA. The image has been broken into a large number of squares called pixels. The computer recognizes the rough outline of the cell nucleus and estimates the optical density of each pixel within the nucleus. The overall density gives a measure of the DNA within that one cell.

The operator has very direct control over the choice of cells to be analyzed in image cytometry. Looking at the slide through the microscope, the operator can select each cell to be analyzed. A small cluster of tumor cells embedded in a benign fibrous stroma can be analyzed. The operator can even choose to look at just the irregularly shaped cells, or just the large cells, or just the cells in one part of the tumor versus another part. The DNA cytogram produced by an image cytometer is similar to those seen from flow cytometry (as in Figs. 4.5 and 4.6). Whereas flow cytometry

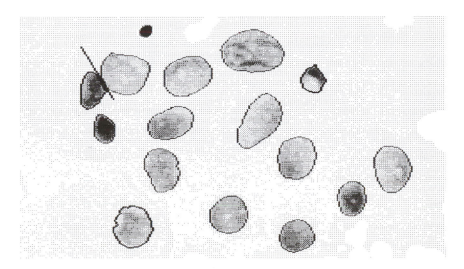

FIGURE 4.6. Image cytometry digital analysis of a group of cells.

typically analyzes 10,000 to 30,000 cells, image cytometry is based on 100 to 300 cells. Thus, while image cytometry is more selective about the population to be analyzed and more sensitive for detecting small, abnormal populations, flow cytometry is more precise in determining the overall DNA ploidy index and the cell cycle parameters.

A clinical molecular diagnostics laboratory usually employs both flow and image cytometry to cover the full spectrum of clinical samples. Image is best on tumor analysis of small-needle biopsy samples; flow cytometry is best for large tumors and immunology studies.

Bibliography

Dressler LG, Bartwo SA (1989) DNA flow cytometry in solid tumors: practical aspects and clinical applications, Semin Diagn Pathol 6:55–82.

Keren DF, Hanson CA, Hurtubise PE (eds) (1994) *Flow Cytometry and Clinical Diagnosis*. American Society of Clinical Pathologists, Chicago.

Melamed MR, Lindmo T, Mendelsohn ML (eds) (1990) *Flow Cytometry and Sorting*, 2nd ed. Wiley-Liss, NY.

Ross DW (1993) Clinical usefulness of DNA ploidy and cell cycle studies. Arch Pathol Lab Med 117:1077.

PART II MOLECULAR APPROACH TO DISEASE

CHAPTER 5

Infectious Diseases

Introduction

In Part I we considered the basic principles and technology that underlie molecular medicine. This chapter on infectious diseases begins Part II in which we apply recombinant DNA technology to the study, diagnosis, and treatment of disease. For infectious diseases, the application of molecular medicine is both straightforward and dramatic. We can take any clinical sample and analyze it for the presence of foreign, nonhuman DNA. The source of the foreign DNA can be identified as to genus and species of the invading infectious agent.

Restriction enzymes are one of the important tools for the identification of microbial DNA from organisms present in a clinical sample. Our clinical application of restriction enzymes is particularly gratifying since bacteria use restriction enzymes in an analogous role. Recall that restriction enzymes are synthesized by bacteria to help them identify and destroy invading viral DNA (Chapter 3). In bacteria, a restriction enzyme recognizes viral DNA sequences and cuts the sequence to inactivate the infectious agent. Our use of restriction enzymes as a tool in diagnosing infectious disease copies the action of bacteria in their defense against viruses. Recently the analogy has been carried one step further, leading to the development of major new potential therapies against infectious diseases. This new therapy is called antisense oligonucleotide therapy. Antisense oligonucleotides are small pieces of DNA that are synthesized to match a portion of the DNA of an infectious agent. The antisense oligonucleotide attaches to the foreign DNA and inactivates the invading microbial organism.

The basic science of applying recombinant DNA technology to infectious diseases is straightforward. We are not looking for an error or alteration in the human genome, as will be the case in Chapter 6, which discuses human genetic diseases. Instead, we are simply looking for DNA sequences of nonhuman origin in places where they should not be, such as inside the sterile environment of the body.

Reflect for a moment on our current methods for the diagnosis and treatment of an infectious disease. A patient presents with signs and symptoms that although nonspecific in nature suggest an infectious process. These initial signs we know to be due in part to the body's inflammatory response and in part to the infectious agent itself. The touchstone of diagnosis is the identification of the infectious agent, which is mostly accomplished by culturing the microorganisms in the laboratory. Broths, petri plates, and biochemical tests are the standard tools of the microbiology lab.

Some infectious agents are easy to demonstrate. Beta hemolytic streptococci from a throat swab are grown overnight on a blood agar plate. Colony counts for *E. coli* in a urine sample are available in 24 hours. *Klebsiella* can frequently be identified in a Gram stain of the sputum. *Mycobacterium tuberculosis* can be seen on an acid-fast bacillus stain of a lymph node biopsy. At other times the infectious organism cannot be easily identified by the microbiology lab and the diagnosis is made clinically, based on the known characteristics of an infection. *Rotavirus* is a cause of diarrheal gastroenteritis, *Rhinovirus* is a cause of an upper respiratory infection, and varicella zoster virus is the agent of chicken pox and shingles. Yet these agents are rarely identified by the clinical laboratory; their presence is presumed based on symptoms. We are accustomed to the usefulness as well as the artifacts and pitfalls of our current clinical and laboratory approach to diagnosis of infectious disease. We know how to interpret the lab results most of the time. A throat culture that the lab reports as mixed flora causes no excitement. A single isolate of *Staphylococcus epidermidis* in a blood culture is usually regarded as a skin contaminant.

Molecular medicine is changing our entire focus on infectious diseases. Before I summarize the current state of the molecular approach to the diagnosis of infectious disease, it is worthwhile to envision what the ideal result of applying recombinant DNA technology could be. The actual is far less than the ideal, but the development of the molecular approach to infectious diseases is so rapid that what we can do today is little measure of what we will be doing next year.

A patient presents in the clinic with mild fever, nasal congestion, discomfort, and cough. A swab of his throat is taken. Instead of culture to identify abnormal microorganisms by their pattern of growth, the sample is analyzed by recombinant DNA techniques. The cotton throat swab is mixed with a cocktail of DNA probes. Enzymes that digest and release the DNA from both host cells and invading bacteria make the DNA in the sample immediately available for hybridization to the probes. The swab is swirled in the liquid mix of the prepackaged test kit for 1 minute. The liquid is then poured through a column that separates hybridized DNA molecules (bacterial target DNA sequences bound to probe DNA) from all other debris. A chemiluminescence detection system for the probes shows two of several possible colors indicating mixed infection. The diagnostic result, available in 10 minutes, indicates a *Rhinovirus* of a strain known to be epidemic in this geographic area. A significant superinfection with a penicillin-resistant streptococcus is also identified. With a definitive diagnosis, the patient is started on the appropriate antibiotic.

Another patient enters the hospital with high fever and intense headaches. A spinal fluid tap is submitted to the molecular microbiology lab. After screening for several viruses, a species of equine encephalitis virus is identified that is endemic to a location recently visited by the patient. A call to the Centers for Disease Control (CDC) results in the emergency delivery of a new antiviral agent. Antisense oligonucleotides are injected into the cerebral spinal fluid. These small DNA pieces bind directly to the virus and block its further proliferation. A temporary reservoir giving access to the cerebrospinal fluid (CSF) is placed and infusion of this therapeutic molecular inhibitor of the virus continues for 5 days until signs of encephalitis have passed.

Neither of the above examples is as yet possible in exactly the scenario presented, but neither are these future applications unimaginable. Technology is already available that makes these examples possible in the near future. The detection of microorganisms by virtue of their unique DNA or RNA components is more direct than identification based on culture methods. Classical descriptive microbiology that identifies bacteria by the morphology of colonies on a petri plate and subsequent biochemical analysis can be replaced by potentially better methods. The description of *E. coli* as a bacterium that forms putrid-smelling colonies on blood agar is an example of a dated technology.

Every species on the earth has a unique genome and each species can be identified by DNA probes. The potential for molecular diagnostics in the identification of microorganisms is virtually unlimited. I believe the etiology of a number of new infectious diseases will be discovered because of the capabilities of recombinant DNA techniques. In this chapter we consider the use of recombinant DNA technology for the detection of microorganisms. The AIDS virus, whose discovery was dependent on recombinant DNA technology, is considered in more detail. The epidemiology of infectious diseases is reexamined in view of the capability of molecular techniques. Finally, the basis for new direct antimolecular therapies, such as given in the fictitious example of equine encephalitis virus, is discussed.

Detection of Microorganisms by DNA Probes

Infectious agents are identified molecularly by recognizing microbial DNA or RNA sequences. A probe is constructed that hybridizes to the microbial target sequences. Hybridization can take place in solution or on a solid medium. The *stringency* of the hybridization reaction is adjusted by varying the salt concentration, buffer, or temperature. Stringency affects the sensitivity and specificity of the binding between probe and target. Probes are labeled to permit visualization and quantitation of their binding to target

DNA. A wide choice of label molecules can be attached to the probe, such as radioactive, chemiluminescent, or enzymatic markers. Table 5.1 lists some of the choices to be considered when designing a probe for detection of an infectious organism. Later in this chapter, I give several specific examples of available commercial kits for the molecular diagnosis of infectious disease. However, this area of molecular medicine, like all the others, is in a rapid state of development. It is more important to understand the potential of recombinant DNA technology for infectious diseases than it is to learn the current available applications, which are still limited.

The PCR is an alternative to direct hybridization methods. The PCR can be used to amplify a small piece of microbial DNA to an amount that is easily detectable. In the PCR technique, a pair of oligonucleotide probes hybridize to either side of the sequence. The target sequence between the probes is copied by cycles of DNA synthesis resulting in an exponential increase in target DNA. The final product, present in a millionfold-increased concentration, is easily detected on a gel or dot blot. The PCR technology greatly increases the sensitivity of detection for microorganisms.

Every microorganism has genomic sequences that make it unique. To develop clinically useful molecular probes for infectious diseases, it is necessary to decide which species of organisms you wish to identify. A probe can be made very specific for a single strain of one bacterial species, or a probe can have broader specificity, encompassing many species. For example, it would be possible to make a probe to hybridize to DNA sequences common to all coliform bacteria. It would also be possible to become much more specific and find DNA sequences that are unique to enterotoxigenic *E. coli*. For various applications such as screening vs. specific subspecies identification, different specificities in the DNA probe are required. For screening, less specificity is desirable. A test to screen for infection should include a wide range of possible organisms, given the symptoms and presenting site of infection. This might require a mix of several probes to identify different species of bacteria and other microorganisms. For very focused applications, such as typing the strain of a *Salmonella* species implicated in an outbreak of food poisoning, it would be desirable to be very specific so that the point source of food poisoning could be determined.

Infectious disease DNA probes can be directed against DNA in the ge-

Table 5.1. Parameters in the molecular detection of microorganisms.

Probe	Hybridization	Detection
Sensitivity vs. specificity	Stringency (strength of chemical attraction between probe and target)	Sensitivity vs. ease and speed
Target sequences to many or single species	Na^{2+}, temperature, buffer	Radioactivity
		Enzyme-lined
Ultrasensitivity PCR	Liquid/solid phase	Chemiluminescence

nome of the microorganism; however, in some instances, other components of the microorganisms serve as better targets. For example, in bacteria, ribosomal RNA, which is present as a large number of copies per bacterium, can be a better target than DNA, which is present in only a single copy. Ribosomal RNA as a target also broadens the range of bacteria that the probes can detect. The choice of target sequences within the infectious agent determines the specificity of the probe.

After the target sequence is chosen, the next technical decision is how to label the probe. For infectious diseases, speed of diagnosis is highly desirable. Toward this end, probes with nonradioactive labels that can be detected quickly without a long photographic exposure have proven useful. Additionally, nonradioactive probes are chemically more stable than their radioactive counterparts and therefore have a longer shelf life. Choice of the probe label, along with the stringency of the hybridization reaction, determines the overall sensitivity of the molecular detection system. For example, chemiluminescent labels are ultrasensitive and can be detected when only a few molecules of probe are bound to a target DNA. In contrast, chromogenic labels, such as enzymes, are less sensitive. A less-sensitive probe can be effective when infection with a relatively heavy large number of bacteria is the common clinical presentation. In many applications, a probe that is ultrasensitive and responds to the presence of only a few bacteria will lead to false-positive reactions.

There are other possibilities for the detection of microorganisms, including hybrid systems that combine standard culture and molecular diagnostics. In some applications, it is better to do a primary culture in which the growth medium inhibits nonpathogenic bacteria while allowing an increase in the target organism. From primary cultures, DNA probes can be used to speed the final identification of the bacteria present. In this application, DNA probes replace complicated biochemical assays.

Table 5.2 (adapted from McGowan, 1989) lists examples of the infectious diseases for which molecular probes already exist. Research is rapidly expanding the number of infectious agents that can be efficiently detected via molecular diagnostics. Adaptation of this research to clinical use is also progressing, but at a slower rate. Firms developing molecular diagnostic kits must balance perceived need with the cost of clinical trials and licensure of their product.

Advantages and Disadvantages of Molecular Detection

The sensitivity of recombinant DNA techniques for the detection of microorganisms is usually very much greater than standard cultures. A PCR method will give a positive result if only a few microorganisms are present. Does a single mycobacterium in the sputum indicate tuberculosis? Experience with adjusting the sensitivity of recombinant DNA techniques to clinically significant levels and interpretation of the data will be required.

Table 5.2. Infectious diseases and molecular probes. [1]

Agent	Sample
Bacteria	
Actinobacillus	Sputum, wound
Bacteroides	Stool, wound
Bordetella pertussis	Nasopharynx
Borrelia burgdorferi (Lyme's disease)	
Campylobacter spp.	Stool
Clostridium sp.	Stool
E. coli	Food, water, stool
Hemophilas sp.	Throat swabs
Legionella sp.	Sputum, colony
Mycobacterium sp.	Colony
Mycoplasma pneumoniae	Throat swab
Neisseria gonorrhoeae	Urethral discharge
Salmonella sp.	Food
Shigella sp.	Stool
Staphylococcus aureus, methicillin resistance	Blood
Yersinia enterocolitica	Food
Viruses	
Cytomegalovirus (CMV)	Urine, blood, lung
Enterovirus	Stool
Epstein-Barr virus (EBV)	Brain, blood
Hepatitis B,C	Liver, blood
Herpes simplex I and II	Tissue, skin
Human immunodeficiency virus (HIV)	Tissue, blood
Papillomavirus	PAP smears, tissue
Parvovirus	Blood, bone marrow
Rotavirus	Rectal swabs, stool
Varicella zoster	Tissue, skin
Parasites	
Chlamydia trachomatis	Eye
Leishmania sp.	Blood
Plasmodium falciparum	Blood
Rickettsia sp.	Blood
Trypanosoma cruzi	Blood
Fungi	
Asgergillus nidulans	Colony
Candida albicans	Colony

[1] Adapted in part from McGowan (1989).

DNA probes, in addition to being too sensitive, also have the potential to be too specific. A probe designed to recognize a single species of a microorganism may not react to other closely related microorganisms. In contrast, culture methods generally show a positive result for a much broader range of microorganisms. For example, a standard blood culture will be positive if virtually any bacteria are present. For comparable DNA detection, it would be useful to develop a mixture of probes that recognizes a broad spectrum of bacteria. Clinicians must be aware of the specificity of the molecular detection system used. Molecular detection systems are a bit like computers in this sense. They will only answer exactly the question you pose. The selectivity of DNA probes is intrinsically very different than the selectivity offered by different growth media used in standard microbial culture.

Recombinant DNA technology is at the present more expensive than standard cultured techniques for most applications. Routine cultures of throat swabs or of urine samples are relatively inexpensive and highly standardized. Recombinant DNA methods would have to be very efficient to compete on a cost basis for some of the most common culture procedures. For organisms that are difficult to grow and identify, DNA methods can be cost-effective. Future automation of DNA technology should significantly reduce the cost of molecular detection systems for the clinical microbiology laboratory.

Notwithstanding the above problems in implementing recombinant DNA technology for the detection of microorganisms, this method has promise because of some inherent advantages. DNA probes for microorganisms can not only identify an organism but can recognize drug-resistant forms. The detection of a penicillin-resistant streptococcus in a throat swab could be made within minutes after the sample was taken. Recombinant DNA techniques work especially well for organisms that are difficult or slow to grow and for viruses and parasites that cannot be cultured at all. For example, mycobacteria can be detected in several hours by recombinant DNA techniques, whereas the standard culture of these organisms can take 3 to 6 weeks to produce a result. For the detection of viruses, not only are recombinant DNA methods less cumbersome and costly than standard methods, but molecular detection can identify a virus in the acute phase of an infection. At the present, a diagnosis of a viral infection is often made by detection of antibodies in the serum of the patient. Antibodies are present only after the infection has persisted long enough for an immune response to occur. The serologic diagnosis of viral infections is not useful in the acute phase of an illness. As antiviral therapies become more widespread, the ability to detect acute-phase viral infections assumes more importance.

DNA probes are particularly valuable for detecting microorganisms in biopsy samples. Special stains for microorganisms—for instance, the Gram's stain—are nonspecific and do not work well in certain tissues. In situ DNA probes can identify the organism with great sensitivity and specificity. The identification of human papilloma viruses (HPV) by in situ hybridization is an important example of this technology (see Chapter 8).

Examples of Molecular Diagnostic Applications

I now turn to the detection of several microorganisms by commercially available kits as examples of how this technology works. I think that you will see that this is an evolving science, still in a relatively primitive phase with respect to clinical applications. I hope that you will also see, however, the potential of recombinant DNA technology for the diagnosis of infectious diseases.

Neisseria

The high prevalence of all sexually transmitted diseases in general, and of gonorrhea in particular, continues to be a source of concern for public health officials and a clinical challenge for physicians. Gen-Probe Incorporated of San Diego has developed a kit for the detection of *Neisseria gonorrhoeae* that uses an ingenious magnetic separation coupled with a chemiluminescent probe. This commercial kit, the PACE-2, has received much attention as an example of what recombinant DNA testing might be able to do for the clinical microbiology lab.

The clinical sample for testing is a swab of the female or male urogenital tract. The swab is inserted into a transport tube and capped. This sample is stable at room temperature for up to 7 days, or much longer if frozen. When the transport tube reaches the lab, the tube is mixed by vortexing, and the swab is removed and discarded. An aliquot of 0.1 mL is pipetted into a hybridization tube. Positive and negative control samples, which are included with the kit, are processed in parallel. Figure 5.1 demonstrates the successive steps that occur in the detection of *N. gonorrhoeae* when this DNA hybridization technique is used. Probe reagent is added to all the sample tubes in the run. The probe reagent lyses the cells and bacterial. The probe solution also contains a DNA probe molecule directed at ribosomal RNA of *N. gonorrhoeae*. The probe is labeled with a chemiluminscent marker. The hybridization tube is incubated for 1 hour at 60°C, allowing stable DNA/RNA hybrid molecules to form. After the incubation with the probe, 1 mL of separation solution is added to each tube. The separation solution precipitates the hybrid DNA/RNA molecules. A second 10-minute incubation at 60°C follows. The tubes are next placed in a magnetized tube holder for 5 minutes. This magnetic separation is directed against a ferromagnetic tag on the DNA probe molecule. The magnet holds back the iron-labeled precipitate. While still holding the tubes in the magnetic separation device, the liquid portion is decanted by inverting the tube rack. The precipitated DNA/RNA hybridization pellet remains in the tube while other molecules in the supernatant are poured off. Each tube is then completely filled with a wash solution and allowed to stand for another 20 minutes in the magnetic separation rack. The supernatant is again poured off by inverting the test tube holder. The small pellet remaining in the tube is resuspended by shaking in the residual drop of wash solution remaining at the bottom of the tube. The tube is now placed in a luminometer, which quantitates the chemiluminscence of each sample. Samples with a luminescence measured as relative light units (RLUs) above a certain threshold value are considered positive for infection with *N. gonorrhoeae*.

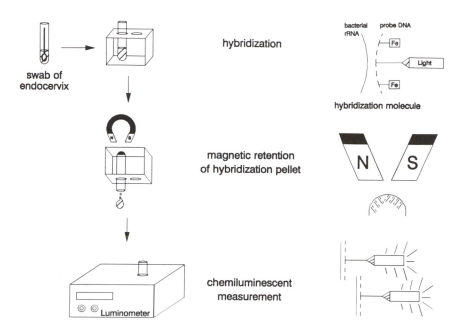

FIGURE 5.1. The detection of *Neisseria gonorrhoeae* by a commercial kit (Gen-Probe, San Diego, CA) uses an ingenious combination of magnetic separation and chemiluminescence to detect hybridization between the DNA probe molecule and target bacteria ribosomal RNA sequences.

Several clinical evaluations of the Gen-Probe PACE-2 system and its earlier developmental stages provide evidence that the results of the DNA test are in agreement with standard culture methods (Granato and Franz, 1989; Panke et al., 1991). The DNA-based test provides a more rapid result, 2 hours vs. 24 to 48 hours for culture. The DNA test does not require rapid transport and processing of clinical samples collected at remote locations, which is necessary for culture. Cost-effectiveness of the PACE-2 kit depends on the volume of tests processed and other factors that are highly variable from lab to lab. A probe for *Chlamydia trachomatis*, another sexually transmitted disease, is also available in the PACE-2 format, and the two tests may be done simultaneously on aliquots from the same clinical sample (Iwen et al., 1991).

Mycobacterium Species

Gen-Probe Incorporated also makes another series of kits called Accu-Probe, which are designed for rapid culture identification. These kits are not yet able to detect infection directly in a clinical sample; however, they can speed up final identification. This is a step toward what must be achieved in a more definitive clinical laboratory test—namely, the direct identification of microorganisms in clinical samples without culture. Some examples of the AccuProbe format are the kits for identifying species of mycobacteria and fungi. Currently probes are available for *Mycobacterium tuberculosis, M. avium*, and *M. intracellulare*, as well as for *Blastomyces dermatitidis, Histoplasma capsulatum, Cryptococcus neoformans*, and *Coccidioides immitis*. After initial culture of the suspected organism from a clinical sample on the appropriate growth medium, a colony is lifted and placed into a hybridization tube. A reagent to lyse the microorganisms and release their nucleic acids is added along with a chemiluminscent probe. A 15-minute incubation at 60°C allows hybrid DNA molecules to form. A second selection reagent is added and the sample is analyzed by using a luminometer to measure the chemiluminescence. The protocol for Accu-Probe is simpler than was the case of direct identification of *Neisseria* in a clinical sample because of the fact that AccuProbe starts with a relatively pure culture of the infectious agent. The necessity of growing colonies from a clinical sample greatly limits the usefulness of this DNA kit for microbial identification. Nevertheless, this application can certainly be seen as evolutionary in recombinant DNA methods. The AccuProbe system is also available with many other probe molecules to confirm the identify of numerous bacterial cultures.

Herpes Simplex Virus

Another company, Enzo Diagnostics, Incorporated (New York, NY) has specialized in testing for viruses by using in situ hybridization methods. Their kits currently come in two formats, called ColorGene and Patho-Gene, available for a number of viruses including herpes simplex virus (HSV), cytomegalovirus (CMV), HPV, adenovirus, EBC, and hepatitis B virus (HBV).

A closer look at the ColorGene kit for HSV will help us understand the in situ method. Herpes simplex virus can cause a number of clinical disorders ranging from fever blisters and genital ulcerative lesions to encephalitis. The severity of the infection and its manifestation depends upon a number of host factors, especially immune status and age. The standard laboratory method for identification of HSV is virus propagation in culture demonstrating a cytopathic effect. The ColorGene kit is designed to confirm the presence of HSV in cultures that already demonstrate a cytopathic effect. The hybridization is carried out in situ on slides prepared of the possibly infected tissue culture cells. The DNA probe for HSV is labeled with biotin. After hybridization, wash steps, and a cytochemical stain for the bound biotin-labeled molecules, the presence of HSV can be visualized within the infected cells. The slides are interpreted by microscopic examination, looking for a brownish precipitate over infected cells. Again, these in situ DNA identification tests are potentially useful, but they still require initial culture. Some research and clinical labs have utilized biotinylated DNA probes for direct in situ detection of virus infections, taking the important step toward a simpler assay (Munoz and Sharon, 1990). These direct in situ tests are not yet widely sold, pending further clinical studies and Food and Drug Administration approval. An exception is the direct detection of HPV in endocervical biopsies and smears, which will be discussed in Chapter 8.

Chlamydia trachomatis

The first of a series of PCR-based diagnostic kits called Amplicor, distributed by Roche Diagnostics (Nutley, NJ), is designed for the detection of *Chlamydia trachomatis*. This organism is a major cause of sexually transmitted disease. Other methods for the detection of *C. trachomatis* include cytology, growth in tissue culture, direct detection in smears with a labeled antibody or DNA probe, and serology. The Amplicor kit has several advantages over all these approaches and serves as a model for the application of PCR to infectious diseases. PCR is sufficiently sensitive to be useful in samples with a low yield of organisms such as urine from asymptomatic males. A PCR-based test has a turnaround time of hours, much faster than most alternatives. The Amplicor kit incorporates a number of technical advances in PCR that make the method suitable for a hospital laboratory. Appropriate positive and negative controls, methods to minimize carryover, a robust amplification protocol, and microwell detection of the amplified product are all part of the kit. Roche has announced further kits to be issued in the Amplicor format for *Enterovirus, M. tuberculosis*, hepatitis C virus (HCV), HIV, and human T-cell lymphoma virus (HTLV)-I and II.

Microbial DNA Testing in Development

The kits now commercially available for the detection of infectious agents are merely the tip of the iceberg and many other kits are becoming available as the technology and demand grows. The development of commercial kits is subject to many other pressures besides availability of the appropriate technology. Manufacturers must consider the cost of complying with government regulations in an area that is still so new that review procedures are certain to change. Additionally, manufacturers must consider the marketability of their product, not only its technical success. Some of the most important applications of DNA probes involve the diagnosis of uncommon diseases, which is a limited market. I will briefly review other applications of DNA technology for the diagnosis of infectious diseases that have not yet been developed into commercial kits. These examples are selected form the hundreds of applications recently published in the research literature. They serve to demonstrate what may be routinely available in the clinical microbiology laboratory within the next few years.

Parvovirus

Human parvovirus B19 infection is associated with sudden shutdown of red cell production—an aplastic crisis—in patients with sickle cell anemia and hemolytic anemia. Parvovirus B19 can also produce an infection of otherwise healthy children that occasionally becomes serious also due to transient suppression of erythropoiesis resulting in an acute anemia. The disorder is called transient erythroblastopoenia of childhood (TEC). If TEC is recognized as the cause of a sudden anemia in a child, only short-term therapy needs to be provided to the patient. The detection of parvovirus B19 infection has been recently accomplished by detecting antibodies to the virus. However, antibodies are present only after the infection and anemia have subsided. Detection at this stage is useful for epidemiologic studies but is too late to be useful in diagnosing the acute phase of the illness. PCR analysis can detect the virus immediately after infection. A 1-day turnaround time for the assay is possible, giving a result that is timely. Initial clinical studies have shown this approach to be both sensitive and specific (Sevell, 1990). The availability of the PCR test makes possible a specific diagnosis of TEC and the distinction from the more dangerous acute aplastic anemia.

Parvovirus B19 may be associated with more diseases than was previously recognized. Using an in situ hybridization assay for the virus in fetal tissues, a German group found a high association between B19 infection and spontaneous abortion (Schwarz et al., 1991). The molecular detection of this virus constitutes a new tool of medical research. It is not surprising that given a new effective tool, clinical studies have begun to find a wider range of diseases resulting from parvovirus B19 infection. As molecular probes of infectious agents continue to become available, this is a story that will be repeated many times for other microorganisms.

Malaria

Plasmodium falciparum, the infectious agent of falciparum malaria, is most commonly detected by examining blood films for the causative organism. The organism is present, usually in large numbers, when the patient's fever is rising or is high; it is rarely seen in the blood between fever spikes. In Third-World countries, the diagnosis of this deadly form of malaria is usually made on clinical grounds, without confirmation by blood film analysis. The development of antimalarial vaccines, however, requires better objective measures of the incidence of a specific strain of infection. A DNA hybridization method for detection of malaria has been shown to be equivalent in sensitivity to the blood smear examination. This molecular probe has been employed in field trials in medically underdeveloped areas (Barker et al., 1989). The DNA method also had less observer bias than diagnosis made by clinical criteria.

Table 5.2 lists a number of infectious diseases for which molecular probes have been adapted to assist in the diagnosis. This table is large but in no sense complete. The infectious agents listed in Table 5.2 are an eclectic mixture of some common and some rare microbes. This is also true for the examples I have just given. Molecular probes have been adapted first to agents that are difficult or impossible to grow, and second to rare disorders or diseases in which the role of the infecting agent is not completely understood. As the technology progresses, the competition between molecular diagnosis and standard culture techniques will become "head to head" regarding common diseases.

AIDS

The acquired immunodeficiency syndrome (AIDS) has been a crucible for recombinant DNA technology. The rise of recombinant DNA technology occurred during the same period that AIDS was discovered. Recombinant DNA methods found HIV, its mode of entry into the cell, its spread via body fluids, and its relationship to other retroviruses. I find it interesting to speculate on how the history of the AIDS epidemic would have differed had this infection occurred 10 or 20 years earlier, prior to the development of recombinant DNA technology. I believe that we would eventually have recognized the syndrome as infectious, based on good epidemiologic evidence. I doubt that we would have known that AIDS was viral in origin. The host of opportunistic infections could easily be (and has been) mistaken as the primary cause of the disease. Fortunately, recombinant DNA technology progressed rapidly during the rise of AIDS, in part due to the impetus from the large-scale research effort surrounding the epidemic.

Life Cycle of HIV

The biology of the HIV-1 retrovirus, the etiologic agent of the human acquired immunodeficiency syndrome, is known in considerable detail. The powerful tools of recombinant DNA technology and the intense level of research have revealed the entire life cycle of the virus, which I have shown schematically in Figure 5.2. Like all retroviruses, HIV uses RNA as its genetic material rather than DNA. The RNA is enveloped in a protein coat. The virus can only infect certain cells to which the viral protein coat can bind. This attachment occurs via a receptor called CD4 (shown at 1 in Fig. 5.2), which is found predominantly on helper T lymphocytes. After the virus binds, the RNA enters the cell and the viral genome takes over the cell's metabolic machinery. The virus makes use of a unique enzyme called *reverse transcriptase* to achieve its takeover of the cell. This enzyme makes a DNA copy of the RNA viral genome (shown at 2 in Fig. 5.2). This act of copying RNA back into DNA is the reverse of the normal flow of information in cells, hence the names *retro* virus for this class of virus and *reverse* transcriptase for this unique enzyme. The virus, after copying itself into a DNA form, directs the cell to make more viral copies, which will in turn be released to propagate the infection. The DNA copy of the viral genome may enter the nucleus of the cell and become integrated into the cell's genetic material (shown at 3 in Fig. 5.2). The infection may, at this point, enter a latent stage with the viral DNA remaining in the infected cell nuclei,

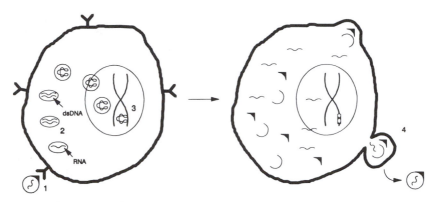

FIGURE 5.2. Life cycle of the human immunodeficiency virus begins with attachment of the virus to a specific receptor on the cell surface (1). The RNA core enters the cell (2) and serves as a template for a DNA copy. The virus replicates as a double-stranded (ds) DNA molecule, eventually becoming integrated into the cell's DNA in the nucleus (3). The cell may remain in this latent phase or may direct reproduction of new viral RNA strands, which are coated with proteins and then released by budding off from the cell surface (4).

but with no new virus particles being produced. After a period of time, the infection can become active again with the viral DNA directing the cell to make the protein and RNA components of the mature viral particle. This is shown in the right half of Figure 5.2. Viral particles when completely assembled bud off from the cell's cytoplasm (as shown at 4 in Fig. 5.2).

The viral RNA genome is small and contains genes for the reverse transcriptase enzyme and for encoding the various proteins of the viral particle coat. Figure 5.3 is a gene map of the HIV virus. The entire neucleotide sequence for the virus is known. Variations in the gene map of HIV allow the identification of different strains of the virus. Since the virus itself contains only a few critical genes, the virus depends on the cell's own machinery. This close dependence of the virus on the cell's metabolism makes it difficult to find drugs that disrupt the virus without damaging the

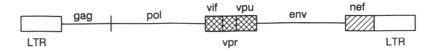

FIGURE 5.3. A genetic map of HIV-1 shows the major structural proteins (gag, pol, env) common to all retroviruses. Regulatory protein (nef) and other genes (vif, vpr, vpu) are also noted. The viral genome is bracketed by long terminal-repeat (LTR) sequences that are involved in regulation of viral replication.

cell. Figure 5.4 shows a schematic diagram of a complete HIV virus particle. The various protein components of the viral coat are labeled with their molecular weights. These viral proteins are the antigens that provoke an antibody response from the patient. The Western blot shown in Figure 3.3 reveals the antibodies commonly found in the blood of an HIV-infected person. This antibody response, however, does not appear to be a sufficient defense against the infection.

Clinical Description of AIDS

The infection begins with entry of the virus into the blood. After a short incubation period during which the virus proliferates, the initial clinical manifestation of AIDS is a nonspecific flulike syndrome, similar to many viral infections. A transient, usually generalized, lymphadenopathy occurs that may last anywhere from a few weeks to months. During the acute phase, virus may be detected, but antibodies to the HIV protein coat have not yet developed. AIDS is infrequently diagnosed in this initial phase.

The acute phase of AIDS resolves with disappearance of the nonspecific clinical symptoms and resolution of the lymphadenopathy. The patient enters a latent phase that typically lasts from 1 to 5 years. The only evidence of AIDS in the latent phase is laboratory evidence; presence of the virus or of antibodies to HIV. These antibodies to HIV proteins are detectable 6 to 12 weeks after the acute infection and generally persist throughout the patient's remaining life.

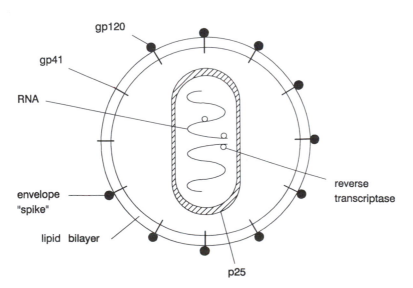

FIGURE 5.4. The structure of an intact HIV viron shows the important envelope and core proteins and central RNA with adjacent reverse transcriptase (RT) enzyme.

During the latent phase, the patient is asymptomatic but still able to transmit the infection. The virus is present in blood, saliva, and semen. Fortunately the HIV virus is exceedingly unstable outside the body. It is easily killed by disinfectants and even by prolonged exposure to air with drying. As the latent phase ends, the patient gradually becomes host to an increasing number of opportunistic infections.

The transition from an asymptomatic latent period to a serious illness is frequently heralded by a phase called AIDS-related complex (ARC). Generalized lymphadenopathy recurs at this stage and the patient is noted to have a depressed lymphocyte count and anemia. A progressive fall in the number of helper T lymphocytes in the peripheral blood can now be measured. This laboratory evidence of impairment to cell-mediated immunity occurs along with clinical signs of infections due to agents such as *Pneumocystis carinii* and *Candida albicans* as the disease progresses. Over a period of months ARC develops into full-blown AIDS. Opportunistic infections, wasting, development of Kaposi's sarcoma, and a severe neuropathy are the main features of terminal AIDS. The longer a patient remains alive, the more likely that a high-grade lymphoma will develop. Lymphomas are seen with increasing frequency as therapy for other aspects of AIDS improves, allowing the patient to live longer. Death from AIDS is most commonly due to infection.

Diagnosis of AIDS

The diagnosis of AIDS is made by detecting either the virus or antibodies to the virus. The virus can be grown in culture, but this is slow, difficult, and requires extreme safety precautions. Detection of the virus by PCR, which can now be done through a commercially available kit, is much more efficient. In this method, the viral RNA, if present in a clinical sample, is amplified to a degree sufficient to produce a visible signal (Dahlen et al., 1991; Ou et al., 1988). Blood or tissue is added to a buffer that releases DNA and RNA from cells. A pair of oligonucleotide primers specific for the HIV-1 genome is added along with other components of the PCR reaction mix. Next, the sample undergoes thermal cycling, each cycle causing a doubling in the amount of HIV DNA present. After 30 cycles, HIV DNA is increased by greater than 1-millionfold. Gel electrophoresis, dot blotting, or fluorescent detection of the PCR product will confirm the presence of HIV. If no virus is present in the initial clinical sample, then the amplification has no template to work from. In that case, no HIV DNA is synthesized, producing a negative result.

The PCR method for detecting HIV is so sensitive that the presence of only a few copies of the virus in the original sample produces a positive result. This appears to be a clinically appropriate level of sensitivity, since any degree of virus infection is association with the development of AIDS. However, clinical labs employing this test must pay exquisite attention to the problem of false positives due to contamination. If only a few molecules of HIV DNA are carried over from a true-positive sample into the start of another test, a false-positive result will be produced. Until laboratories became aware of the special problems in preventing even the most minute cross-contamination in PCR testing, false positives were the bane of this technique. For a period of time, the contamination problem almost ended PCR-based testing, and the results are still considered somewhat skeptically. Some very clever techniques to prevent cross-contamination have now been developed, and PCR testing is valid if carried out by a competent laboratory.

The detection of HIV virus by PCR is usually reserved for a few specific applications. The diagnosis of AIDS is normally made by detecting the presence of antibodies to HIV in the serum. HIV antibodies are usually detected by a screening enzyme-linked immunosorbent assay (ELISA) procedure and then confirmed by Western blot. However, the presence of the HIV viral genome in the blood can be detected by PCR immediately after infection begins, whereas antibodies do not develop for 6 to 12 weeks. Thus PCR as a diagnostic technique can be used to fill in the "window" between infection and the development of antibodies. Detecting the virus may become more important if therapeutic measures for interfering with the acute infection are discovered.

Therapy of AIDS

The antiviral treatment of AIDS can be directed against several points in the virus's life cycle (as shown in Fig. 5.2). Artificial soluble CD4 molecules can be introduced into the blood to confuse the virus so that it misses its target on the surface of a cell. Clinical trials of soluble CD4 for AIDS have begun.

The next step in the viral life cycle where drugs have been shown to be effective is where they can interfere with reverse transcriptase. The successful inhibition of reverse transcriptase by AZT (zidovudine, Burroughs Wellcome), and by the similar experimental drugs dideoxyinosine (ddI) and dideoxycytidine (ddC), is the current best treatment of AIDS. Other drugs that interfere with reverse transcriptase and have lower toxicity (such as compound BI-RG-587) are also promising.

Other means of interrupting the virus's life cycle after the viral RNA has been copied to DNA are under investigation. Specific proteases may destroy viral packaging of the protein coat. Vaccines that induce antibodies to the viral proteins may offer protection for noninfected persons at risk. The development of a vaccine to establish protective antibodies in noninfected people depends on producing a genetically engineered virus with a protein coat, which is a strong antigen without the danger of infection. The ability of recombinant DNA genetic engineering to manipulate the virus and isolate specific portions of its genome that code for surface envelope proteins is the major goal in vaccine research. The globular proteins in the viral protein coat shown in Figure 5.4 are the current targets for development of a vaccine.

Epidemiology of Infectious Diseases

Molecular probes, because of their potential for exacting specificity, are useful to epidemiologists in the study of the spread for infectious diseases. The route of transmission of an infectious agent is one of the things to understand in preventing disease. Molecular probes for infectious agents can potentially identify clones of a single organism. This allows microbial isolates from various sources to be compared for identity.

The study of an outbreak of a diarrheal illness due to food contamination serves as an example of the potential of molecular probes to work as a tool in infectious disease epidemiology. *Salmonella* is an important agent in the cause of gastroenteritis and is frequently transmitted by contaminated food. In an epidemic where this organism has been isolated from stools of a number of individuals, it would be important to know if all the isolates are identical, indicating a single source for the infection. In the past, careful epidemiological studies based on interviews with patients and tracing their contacts has been the major method for implicating a point source as the cause of a food-related outbreak of gastroenteritis. Molecular probes for an individual clone of *Salmonella* can significantly aid this process (Rivera et al., 1991).* The bacteria present in multiple patients can be compared to samples obtained from potential food sources. The important question of whether the outbreak is the result of a single source such as an infected food handler or represents poor food-processing hygiene resulting in multiple contaminated chicken carcasses can be determined by this epidemiological approach.

Molecular probes also serve to genotype specific strains of bacteria that are virulent against a background of nonpathogenic species. For example, outbreaks of enterotoxogenic hemorrhagic *E. coli* must be detected against the background of the ubiquitous presence of *E. coli* organisms in the stool. The standard approach for characterizing substrains of an organism is to phenotype an isolate. Methods for phenotyping of bacteria include serotyping, bacteriophage susceptibility testing, and antimicrobial susceptibility patterns. For *E. coli*, serotyping has been the most useful; the strains with serogroups O157:H7 and O26:H11 have been associated with hemorrhagic enteritis. The O and H antigens are not the primary cause of virulence in this strain of *E. coli*; they are phenotypic markers of the virulent strain. Since DNA probes can be exquisitely specific, the potential to replace complex phenotyping of bacteria by more technically simple DNA probes

---

*In Chapter 6, we will see how DNA fingerprinting can identify individuals. The use of probes sufficiently specific as to be able to identify individual clones of bacteria is in reality the same process of DNA fingerprinting, applied to bacteria.

exists. This potential is now being realized but is still more theoretical than actual.

Antisense Oligonucleotides as Antimicrobial Agents

In Chapter 2, the theory of antisense oligonucleotide inhibition of gene expression was introduced. We saw that short pieces of DNA (oligonucleotides) synthesized to be a mirror or antisense match to portion of an mRNA molecule could block the translation of RNA into protein. One of the first and most successful applications of antisense technology is to interfere with viral replication within a cell. Antisense oligonucleotides specific to a viral genome will block the virus — either its expression or its replication within a cell. Viruses that have been blocked by this method include HSV and HIV. Antisense oligonucleotides constitute a new type of antiviral drug with a high degree of specificity. Experimental tests in cell culture and in animal systems have already shown marked inhibition of viral infection in the presence of therapeutic levels of the specific antiviral antisense oligonucleotide. Other microorganisms larger than viruses can also be inhibited by antisense oligonucleotides as has been demonstrated for *Trypanosoma brucei* (African sleeping sickness).

New Infectious Diseases

One tends to think of newly discovered infectious diseases, but not of truly new organisms. For most of the plants and animals on the planet, we tend to think in terms of extinction with loss of species. Microbial agents are different. They mutate rapidly. A generation for a virus or a bacterium can be 20 minutes. The HIV virus is probably a relatively new organism, mutated from a simian virus in Africa. HIV appeared in the world because of the much greater spread of people resulting from rapid international travel. As pointed out in the documentary book *The Hot Zone* (Preston, 1994), a virus is within a 24-hour plane trip of any point on the earth. Recombinant DNA technology increases the possibility for new organisms, both beneficial and harmful. DNA engineering has successfully turned bacteria and yeast into new species that produce human drugs such as insulin. One can imagine many bioengineered microorganisms serving multiple roles in the environment, from production of drugs to protection of crops. An infectious vaccine is another possibility. If you create an attenuated live virus that stimulates antibody production against a virulent organism, and then spread the live virus at airports and other transportation centers, you will have immunized the world. Imagine the potential for a vaccine against malaria, spread by the very same mosquitoes that cause the disease.

Unfortunately, recombinant DNA technology can also be used to create harmful infectious agents. Biological warfare or terrorism is more feasible because the means to manipulate microorganisms has been so greatly facilitated. It may take a very large national project to create a nuclear weapon, but a small laboratory buying off-the-shelf supplies can carry out bioengineering of microorganisms.

Conclusions

The application of recombinant DNA technology to the study, diagnosis, and treatment of infectious diseases has just begun. The DNA-probe–based diagnostic kits that are now available for the clinical microbiology lab may reveal how physicians will use and adapt to this technology. Much more can be done to improve diagnosis, both in terms of accuracy and in cost benefit. Some applications await the improved automation of recombinant DNA technology before they can truly be efficient. DNA probes do offer the opportunity to identify in a routine manner organisms that are difficult to grow. The diagnosis of viral diseases could become as specific and as routine as our current ability to identify bacteria. The precise identification of strains of microbial organisms should improve our knowledge of the epidemiology of infectious diseases. More specific information on which strains of a particular infectious agent are virulent may explain some of the clinical variability seen in infections. The treatment of infectious diseases by antisense oligonucleotides offers some of the greatest excitement within this new technology. It is still too soon to know whether antisense oligonucleotides will be a powerful form of antiviral therapy. The concept of ultraprecise targeting of a specific nucleotide sequence as the site for interference by a drug is very attractive. Antisense may hold a great future, or it may fail for as-yet-unappreciated technical reasons. The fusion of classical microbiology with the new discoveries and technology of molecular biology will, in my opinion, proceed rapidly. Microbiologists are more accustomed to the concepts of molecular biology than most other scientists and physicians. Recombinant DNA technology has been a tool for them longer than for most of us. The very nature of an infectious agent begs for its identification and attack at the level of DNA.

Bibliography

Barker RH, Brandling-Bennett AD, Koech DK, Mugambi M, Khan B, David R, David JR, Wirth DF (1989) *Plasmodium falciparum*: DNA probe diagnosis of malaria in Kenya. Exp Parasitol 69:226–233.

Dahlen PO, Iitia AJ, Skagius G, Frostell A, Nunn MF, Kwiatkowski M (1991) Detection of human immunodeficiency virus type 1 by using the polymerase chain reaction and a time-resolved fluorescence-based hybridization assay. J Clin Microbiol 29:798–804.

Eisenstein BI (1990) New molecular techniques for microbial epidemiology and the diagnosis of infectious diseases. J Infect Dis 161:595–602.

Figueroa ME, Rasheed S (1991) Molecular pathology and diagnosis of infectious diseases. Am J Clin Pathol (Suppl) 95:S8–S21.

Granato PA, Franz MR (1989) Evaluation of a prototype DNA probe test for the noncultural diagnosis of gonorrhea. J Clin Microbiol 27:632–635.

Hillyard DR (1994) The molecular approach to microbial diagnosis. Am J Clin Pathol 101 (Suppl 1):518–521.

Iwen PC, Tina MS, Blair MH, Woods GL (1991) Comparison of the Gen-Probe PACE 2TM system, direct fluorescent-antibody, and cell culture for detecting *Chlamydia trachomatis* in cervical specimens. Am J Clin Pathol 95:578–582.

McGowan KL (1989) Infectious diseases: diagnosis utilizing DNA probes. Clin Pediatr (Phila) 28:157–162.

Ou CY, Kwok S, Mitchell SW, Mack DH, Sninsky JJ, Krebs JW, Feorino P, Warfield D, Schochetman G (1988) DNA amplification for direct detection of HIV-1 in DNA of peripheral blood mononuclear cells. Science 239:295–297.

Panke ES, Yang LI, Leist PA, Magevney P, Fry RJ, Lee RF (1991) Comparison of Gen-Probe DNA probe test and culture for the detection of *Neisseria gonorrhoeae* in endocervical specimens. J Clin Microbiol 29:883–888.

Persing DH, Smith TF, Tenover FC, White TJ (eds) (1993) *Diagnostic Molecular Microbiology. Principles and Applications*. American Society for Microbiology. Washington, DC.

Preston R (1994) *The Hot Zone*. Random House, NY.

Rivera MJ, Rivera N, Castillo J, Rubio MC, Gomez-Lus R (1991) Molecular and epidemiological study of *Salmonella* clinical isolates. J Clin Microbiol 29:927–932.

Schwarz TF, Nerlich A, Hottentrager B, Jager G, Wiest I, Kantimm S, Rogendorf H, Schultz M, Gloning K, Schramm T, Holzgreve W, Roggendorf M (1991) Parvovirus B19 infection of the fetus: histology and in situ hybridization. Am J Clin Pathol 96:121–126.

Sevell JS (1990) Detection of parvovirus B19 by dot-blot and polymerase chain reaction. Mol Cell Probes 4:237–246.

Sloand EM, Pitt E, Chiarello RJ, Nemo GJ (1991) HIV testing, state of the art. JAMA 266:2861–2866.

Tenover FC (1988) Diagnostic deoxyribonucleic acid probes for infectious diseases. Clin Microbiol Rev 1:82–101.

Woods GL (1994) Tuberculosis, role of the clinical laboratory in providing rapid diagnosis and assessment of disease activity. Am J Clin Pathol 101:679–680.

CHAPTER 6

Genetic Diseases

Introduction

The application of recombinant DNA technology to human genetics has advanced our knowledge of inherited diseases immensely. Classical genetics was very limited in its ability to probe the complexity of the human genome. The famous genetic breeding experiments in viruses, bacteria, yeast, slime mold, and fruit flies taught us a great deal about genetics in general. However, applying classical genetics to human diseases was limited by our scant knowledge of human genes. Some people doubted that the science of human genetics would ever make much progress.* Recombinant DNA technology and genetic mapping have opened the very complex human genome to direct study.

In this chapter, the contributions of recombinant DNA technology to the study of human inherited diseases are presented. Too much is happening in this area to attempt to be all inclusive. A brief review of basic genetics will serve as an introduction,** followed by examples of molecular medicine applied to genetic diseases. Table 6.1 is a list of selected hereditary disorders for which details of molecular genetics have been discovered. From this list, cystic fibrosis, sickle cell anemia, and familial hypercholesterolemia are chosen to demonstrate the science as well as the clinical problems associated with molecular genetics. The impact of recombinant DNA technology and the problems posed, both medically and ethically, by this technology, are also discussed.

---

*In the early 1970s, a famous geneticist in Berkeley, California stated that the genetics of humans would likely remain impenetrable to scientific research, and consequently he was leading his research group into the area of neurosciences. The very year this announcement was made, restriction enzymes were first being employed as a means of mapping DNA, across the bay from Berkeley in San Francisco and Palo Alto. Thus, even while this then-famous researcher was speaking, his opinion was being proved wrong.

**For more information, the reader should consult the excellent text by Gelehrter and Collins (1990) listed in the bibliography of this chapter.

Table 6.1. Hereditary diseases and DNA probes.

Disease	
Acute intermittent porphyria (AIP)	Hemophilia A (F8C deficiency)
Adenosine deaminase (ADA) deficiency	Hemophilia B (F9 deficiency)
Alpha-1 antitrypsin (AAT) deficiency	Hereditary breast/ovarian cancer
Alzheimer's disease	Hereditary congenital hypothyroidism
Antithrombin III (ATIII) deficiency	Hereditary fructose intolerance
ApoE deficiency	Hereditary non polyposis colon cancer
Becker muscular dystrophy	Hereditary persistence of fetal hemoglobin
C2 deficiency	Homocystinuria
Charcot-Marie-Tooth disease	Huntington's disease
Chronic granulonatous disease	Insulin resistance type A
Color blindness	Lesch-Nyhan syndrome
Congenital adrenal hyperplasia	Maple syrup urine disease
*Cystic fibrosis (CF)	McArdle's disease
Diabetes mellitus due to abnormal insulins	Mucopolysaccharidosis VII
Duchenne muscular dystrophy (DMD)	Multiple endocrine neoplasia II
Dysfibrinogenemia	Non–insulin-dependent diabetes mellitus
Ehlers-Danlos syndrome	(NIDDM, type II diabetes)
Elliptocytosis-2, spherocytosis	Osteogenesis imperfecta
Encephalomyopathy, mitochondrial	Phenylketonuria (PKU)
Fabry disease	Porphyria cutanea tarda
Factor X deficiency	Retinitis pigmentosa
Familial adenomatous polyposis	Retinoblastoma
*Familial hypercholesterolemia	*Sickle cell anemia
Fructose intolerance	Tay-Sachs disease
Glucose-6-phosphate dehydrogenase (G6PD)	Thalassemia beta (HBB)
deficiency	von Willebrand's disease (VWF) type IIA
Gaucher's disease	

*See discussion in text.

Basic Genetics

Mendelian Inheritance

Long before the molecular structure of the gene was discovered, the basic behavior of genes had been determined from breeding experiments. Animal breeders knew that certain traits would be faithfully represented in the next generation, whereas others might skip a generation and thereafter be only randomly present. Basic genetics starts with an understanding of Mendelian inheritance. A genetic *locus* is a specific position on a chromosome, which we can think of as a slot that holds a gene. Filling that slot is an *allele*. An allele is one of several possible alternate forms of a gene. Since there are two of every chromosome (except X and Y in males), there are at least two loci, and thus at least two copies for every gene. If the two alleles present on each of the chromosomes are identical, the individual is said to be *homozy-*

gous. If the alleles differ, the individual is said to be *heterozygous*. If the presence of an abnormal allele causes the individual to have a disease, we say that the allele has a *mutation*. If an unusual allele does not cause any abnormality, we call the alternate form of the allele a *polymorphism*. The mutant allele may be either dominant or codominant or recessive. Marfan's syndrome is an example of a mutation that is dominant. If a person receives one mutant allele from either parent, some features of the syndrome will be expressed. Familial hypercholesterolemia is a codominant mutation. Individuals who receive one copy of the gene from either parent will have a serum cholesterol that is two to three times normal, whereas people homozygous for the mutation will have a five- to sixfold elevation with serious sequelae.

Figure 6.1 is a pedigree of a family with hypercholesterolemia. The ages of each family member are given, for deceased members (marked with a slash) the age at the time of death is noted. Two of the great-grandparents died relatively young of atherosclerotic disease and are inferred to have had hypercholesterolemia. Therefore, the symbols denoting them are shaded. All three children of this couple died of atherosclerotic disease, one of them of very advanced disease at age 31. That individual may have been a homozygote, and I have marked him with a solid square to indicate this. The subsequent mating of one of the three children of this affected couple

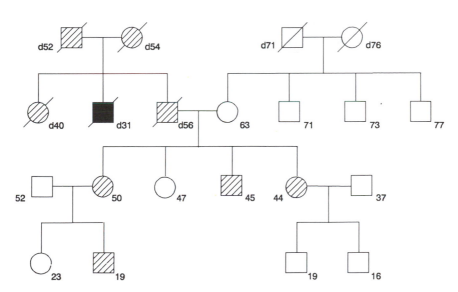

FIGURE 6.1. A pedigree of an autosomal dominant trait, familial hypercholesterolemia, spans four generations of affected individuals. Shaded symbols indicate heterozygotes. The single solid square denotes a presumed homozygote who died at age 31 of atherosclerotic disease. A slash indicates death.

with a person with no family history of hypercholesterolemia resulted in four children, three of whom are affected. Only one of the children of this third generation has passed the inherited disorder on to a son.

Tay-Sachs disease is a recessive mutation. The disease is clinically present only when an individual has inherited mutant genes from both parents. If an individual has one normal gene and one mutant, the single normal copy is sufficient to generate enough of the enzyme (hexosaminidase A) to prevent the accumulation of ganglioside. A heterozygote for Tay-Sachs disease does have decreased enzyme function, but what remains is enough to prevent any manifestation of the disease. Thus an individual with one normal and one mutant gene, a heterozygote, would be phenotypically and clinically normal. A heterozygote of a recessive mutation is also called a *carrier* of the disease. Figure 6.2 demonstrates the pedigree of a family carrying the Tay-Sachs disease trait. One child, indicated by the arrowed, crisscrossed square, has the disease. I have marked the parents of the affected child with half-shaded symbols to indicate that they are carriers. This must be so since they produced a son with the disease. The parents are not clinically affected, because Tay-Sachs is a recessive disease. Compare Figure 6.2 with Figure 6.1, in which heterozygotes are marked with a full shaded symbol rather than a half. This is done by convention to indicate that these heterozygotes have the disease, because familial hypercholesterolemia is a dominant mutation. The pedigree in Figure 6.2 reflects the *phenotype* of the individuals—that is, whether or not they have a clinical abnormality. This family, now that they know that they are carriers of the disease, may wish to seek genetic counseling and further testing to construct a pedigree that reflects the *genotype* of all individuals. The aunts and uncles of the affected

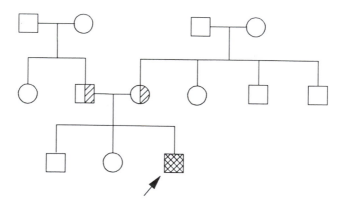

FIGURE 6.2. A pedigree of an autosomal recessive trait, Tay-Sachs disease, shows an affected child (arrow), also called the proband. His parents must be carriers; the gene status of his aunts, uncles, and grandparents is unknown.

child could be carriers of Tay-Sachs disease. Later in this chapter we discuss how molecular probes make it possible to determine the genotype of an individual in this type of clinical situation. The pedigree in Figure 6.2 demonstrates an example of an autosomal recessive disease inheritance pattern.

For diseases in which the gene locus is present on the X chromosome, inheritance and development of disease do not follow exactly the simple Mendelian pattern I have just described. The incidence of sex-linked genetic diseases is much greater in males than in females. This is because males have only one copy of the X chromosome, and if they inherit a mutant gene from their mother, there is no possibility of getting a correct gene from their father, who can contribute only a Y chromosome. Many sex-linked diseases, such as hemophilia, occur almost exclusively in males. The Y chromosome, which is present in a single copy in males and absent in females, has very few genetic loci and appears to consist mostly of useless DNA sequences. The Y chromosome could not have many major functions since no genes present on the Y chromosome can contribute to the biology of female organisms. Furthermore, since males possess only one copy of the Y chromosome, any mutations would always be expressed, frequently resulting in death of the male. Since there is no possible way to have a lethal Y chromosome mutation maintained in a carrier state (there being no YY heterozygote organisms), these mutations disappear from the population. Over an evolutionary time span, the genetic loci (except for sex determination) of the Y chromosome have probably been eliminated by this mechanism.

Most gene systems are more complex than just one locus per chromosome with only one common allele. There can be multiple allels for a single locus, or the function of a protein can depend on multiple loci. The determination of human blood types is an example of multiple alleles at a single locus. In the ABO gene system, an individual can possess one of three common alleles, A, B, or O, at each locus. If the patient is homozygous AA or BB, the blood type is A or B, respectively. If the individual is heterozygous for A and B alleles, then the blood type is neither A nor B but AB. In this instance, both alleles are expressed and the proteins derived from these genes are expressed as antigens on the red cell surface. The AB blood type is an example of codominant inheritance in a multiple allele system. If an individual is heterozygous for A and O (or B and O) the O phenotype is silent, and despite being a heterozygote, the individual demonstrates the A or B phenotype. This is a genetic system with three common allels per locus. Since the various different blood types are not diseases, these multiple alleles are also called polymorphisms.

The human hemoglobin gene system consists of both multiple alleles as well as multiple loci. A hemoglobin molecule in an adult is made of two alpha chains and two beta chains. There are multiple alleles for the two major loci of the beta chain and for the four loci of the alpha chain.

Linus Pauling discovered the abnormal chemistry and inferred a genetic cause for sickle cell anemia in 1949. In his paper "Sickle Cell Anemia, a Molecular Disease" (Pauling et al., 1949) he stated:

> This investigation reveals, therefore, a clear case of a change produced in a protein molecule by an allelic change in a single gene involved in synthesis.

Pauling not only discovered the cause of sickle cell anemia but created the very concept of a molecular disease.

Sickle cell anemia is an autosomal recessive disease caused by both beta chains having the same mutant allele, $beta_S$. A person with the $beta_S$ mutation at one locus and a normal beta gene at the other locus would have sickle cell trait. The hemoglobin produced in this individual consists of hemoglobin A and hemoglobin S. The overall concentration of hemoglobin S would be less than 50% and this is dilute enough to prevent the sickling phenomenon. Patients with hemoglobin S trait rarely express any disease characteristics. There are other possible mutant allels at the beta chain locus besides $beta_S$. An individual can be homozygous for $beta_C$, which gives the phenotype of hemoglobin C disease. A person can be a compound heterozygote (having two different mutant alleles) for both $beta_S$ and $beta_C$. The result is hemoglobin SC, which produces a less-severe clinical disorder than hemoglobin SS because it does not undergo the physical chemical reaction of sickling at low oxygen tensions. In the end, it is the behavior of the abnormal hemoglobin molecule that determines the clinical characteristics of the disease.

Making the system more complex, there are regulatory molecules that control the expression of hemoglobin chains. Beta thalassemia mutations decrease the expression of the beta hemoglobin gene. This is a mutation in a portion of the gene that controls gene expression rather than protein structure. While there are only two loci for the hemoglobin beta chain, there are four loci for the alpha chain. An alpha thalassemia mutation at one of the four loci produces no phenotypic effect, because there are three other loci capable of producing alpha chains. Human hemoglobin synthesis is an example of a complex gene system with multiple loci and alleles. There are other gene systems with multiple loci and very many alleles. The overall behavior of such multigenic systems is quite complex due to interaction of all the various possibilities.

Crossing Over and Meiosis

In addition to the possible complexities of Mendelian inheritance including gene systems with multiple loci and alleles, there is another critically important genetic mechanism called *crossing over*. This is a genetic phenomenon that occurs only at meiosis during formation of the egg and sperm. During meiosis, the two copies of each chromosome come physically close together. This is necessary in order that a haploid gamete may be formed having one copy of each chromosome. However, when two identical chromosomes come close together at meiosis there is the possibility for crossing over. The distal ends of the chromosomal arms are physically exchanged. Genes that were on the same chromosome before crossing over are exchanged so they are now on opposite copies of the chromosome after meiosis. Crossing over is not a pathological process. Its purpose is to increase genetic diversity. Because of this phenomenon, genes can segregate in a fashion that is contrary to Mendelian inheritance. Genes are reshuffled by crossing over at meiosis so that new genetic combinations are created for each chromosome. When two genes are further apart along the chromosome, the chance of them being separated by crossing over increases. Geneticists use a measure of distance between genes called the centiMorgan (cM), which measures the probability of crossing over. Genes that are 1 cM apart along a chromosome have a 1% chance of segregating independently at meiosis due to crossing over. At 50 cM in distance, crossing over is so common that genes separated by this distance behave as if they were on different chromosomes.

This phenomenon of crossing over occurs only once for an organism, during the formation of sexual gametes for reproduction. After the chromosome is reshuffled during meiosis, the haploid egg from the mother and haploid sperm from the father come together and reconstitute a diploid fetus that starts as a single cell. As the fetus grows and proliferates, and in fact throughout the entire life of the organism, crossing over between identical chromosomes does not happen. This is a phenomenon that occurs only between generations.

As two genes get closer together on a chromosome, the chance that they will be separated by crossing over at meiosis decreases. The two genes are said to be *linked*. When two genes are at loci on different chromosomes, or greater than 50 cM apart on the same chromosome, then they sort randomly at meiosis and there is no linkage. The phenomenon of linkage has always been very important to clinical genetics. Very often, we do not have sufficient information about a gene system to allow us to detect carrier status in a prospective parent or to make a prenatal diagnosis in a fetus. However, if the gene causing a disease is linked to some genetic marker we do know about, then we can usually say, if a person has the genetic marker, they will have the disease.

Restriction-Fragment-Length Polymorphism

One of the major molecular techniques for the detection of genetic disease either in prospective parents or in fetal tissues is the *restriction-fragment-length polymorphism* (RFLP) method employing Southern blot hybridization. A genetic polymorphism is an incidental variation in the genome, usually occurring in the noncoding spaces between genes, that has no effect on gene expression. A polymorphism becomes an RFLP when the change in DNA sequence causes a restriction enzyme "cut" site to appear or disappear. DNA from an individual who possesses the polymorphism can thus be recognized on a Southern blot by an altered number of bands after a restriction enzyme digest. An RFLP serves as a useful genetic marker that can permit us to trace inheritance of nearby genes. As an example of genetic polymorphism and RFLP analysis let us consider the problem of DNA fingerprinting, a new application in forensic medicine.

DNA Fingerprinting

One of the major problems in forensic medicine is to establish with certainty the identity of persons, whether alive or dead. Fingerprinting has long been one of the primary means of human identification. Recombinant DNA testing has the potential to replace the conventional fingerprint and offers numerous advantages. Just like the fingerprint, the human genome is different for every individual (except for identical twins). Of the 6 billion base pairs making up the diploid DNA content of a human cell, one person differs from another by about 3 million base pairs, or 0.05%. A few of the differences between individuals are due to mutations, changes within a gene that affect the phenotype. Most of the changes, however, do not occur within the coding portions of the genome, and do not affect the phenotype of the individual. These alterations are called polymorphisms, as was discussed in the previous section. Many of the polymorphisms that constitute individual differences occur in "spacer DNA." These are short segments of DNA that repeat a variable number of times. When a sperm or an egg is formed by the process of meiosis, the number of spacer segments in the germ cell changes in a random fashion. A fertilized zygote is formed at the moment of conception by combining the two haploid germ cells to give a diploid cell. The diploid zygote has different numbers of spacer elements than either parent. As we shall see, this forms the basis for the most common method of DNA fingerprinting.

The human genome is recorded in every cell of the body, not just on the tips of the fingers. All tissues, including blood, saliva, hair, semen, and skin, serve as possible sources of evidence for DNA fingerprinting. Wearing gloves during a crime will prevent fingerprints from being left as evidence, but it is much more difficult to avoid leaving any traces of DNA at a crime scene. DNA is very stable, unlike fingerprints. A dried blood sample may be tested years later for identification.

Several recombinant DNA methods are useful for DNA fingerprinting employing both the Southern blot technique and the polymerase chain reaction. One technique is based on the phenomenon of spacer DNA segments. One type of spacer DNA is called variable number of tandem repeats (VN-TRs). VNTRs are typically 6 to 8 base pairs long and repeat from 6 to 20 times. The purpose of these short, repeating sequences and the number of repeats is unknown. In fact, there may be no purpose to the VNTR phenomenon at all. Nevertheless, in each individual the apparently random repeats of these sequences scattered throughout the genome offers the equivalent of a unique serial number or "bar code" for identifying an individual. Figure 6.3 shows schematically how mapping the pattern of VNTRs in two individuals can lead to a DNA fingerprint. In this example, three different regions (denoted A, B, and C) with VNTRs are examined. A restriction digest for enzyme cut sites on either side of the VNTR (denoted by vertical arrows) releases the DNA fragments containing the VNTR. More copies of the tandem repeats result in a larger DNA fragment. For region A, patient I has 6 repeats, while patient II has only 2. For regions B and C the relative lengths of the VNTR fragments are 5 and 2 for patient I, while for patient II they are 3 and 8. Southern blot analysis of the restriction digest electrophoresis gel probed for the VNTR sequences sorts out the band sizes for each patient as shown in the right half of the figure. The example in Figure 6.3 sows analysis for only three regions. In actual practice, more probes and regions are used to obtain much more specificity, resulting in a blot with very many bands. An actual VNTR analysis looks and functions very much like a bar code, providing a unique signature for DNA from a given individual.

The polymerase chain reaction technique can be used to detect specific site differences in DNA, which will form a DNA signature. PCR methods have frequently employed variations in the HLA-Dq locus of the human histocompatibility gene system. The advantage of PCR is its ability to use very small amounts of DNA, typically less than 0.1 μg, where Southern blot analysis requires 10 μg. Furthermore, DNA even if heavily degraded by chemicals or other factors, is still suitable for PCR analysis.

DNA fingerprinting is offered by several companies and hospital laboratories for a variety of applications. Violent crime, particularly rape, is a major area where semen and blood samples left on the victim can lead to an

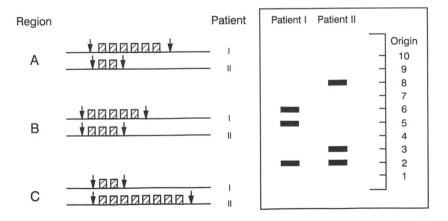

FIGURE 6.3. A schematic diagram of the variable number of tandem repeats (VNTR) shows this method for establishing a DNA fingerprint.

identification of the offender. The complexities of DNA fingerprints as evidence have caught the public attention in the events surrounding the O. J. Simpson trial. Although the technical aspects of DNA analysis are in my opinion well under control, the use of DNA fingerprints is still controversial. Regrettably in the first instance of DNA testing in forensics there was poor technique on the part of some new companies in this field, resulting in erroneous results. This lapse in technique cast some doubt on molecular forensics. Now the F.B.I. and many state agencies are establishing their own DNA laboratories to insure good technique and appropriate control of the chain of evidence.

It will still take some time for DNA fingerprinting to take hold as experience and familiarity with the method by juries is necessary. Whereas everyone knows that wearing gloves will prevent conventional fingerprints from being left at a crime scene, the "tricks" that can confound the DNA fingerprint are not completely know. One company doing paternity testing found that its employees commissioned to draw court-ordered blood samples from prospective fathers were offering to substitute false blood samples for a fee. Other artifacts of DNA fingerprinting are less obvious. While it is difficult to plant someone else's fingerprints at a crime scene, it is simple to leave false DNA. Blood, saliva, or semen collected from some unsuspecting person weeks in advance of a planned crime can be planted at a crime scene. It is even possible to confound a DNA fingerprint by "masking" with pooled DNA. For example, some cosmetics contain human DNA, which if used as a shampoo, might confound a PCR analysis of human hairs found at a crime scene. Nevertheless, the DNA fingerprint when the analysis is done correctly and the evidence is used with good judgment has much to offer forensic sciences.

The DNA fingerprint method has many applications besides crime. Positive identification of victims of accident, catastrophe, and war is far better carried out with DNA than any other method. The U.S. Army is moving toward a DNA based "dog tag" for identification of all soldiers. DNA fingerprinting can also be used for sample identification in the clinical labs. Every blood and tissue sample carries its own internal bar code. For the moment, the use of DNA fingerprinting in clinical samples is reserved to resolve possible sample mix-ups. A prostate biopsy that demonstrates cancer but is misidentified due to poor or lost labeling can be used as a source of DNA to match with a blood sample from the prospective patient.

The future of the DNA fingerprint is uncertain because of the vast potential of DNA technology. In some futuristic time we will all be known by our DNA bar code rather than driver's license, social security number, or credit card number. Our identity will be difficult to hide, and perhaps our privacy infringed upon.

Hemophilia

Hemophilia, a sex-linked mutation causing a defect in coagulation, is another example of an RFLP analysis. Our patient is a male fetus at risk for

hemophilia because the disease has occurred in two of his mother's broth-
ers. Hemophilia results from one of many possible mutations in the factor
VIII gene on the X chromosome. The mother has two X chromosomes, one
of which may carry the hemophilia mutation. The fetus has received one of
the mother's two X chromosomes. The fetus, being male, has not received
an X chromosome from his father. Thus if a mutation is present on the
maternally inherited X chromosome, it will be expressed. We need to know
which one of the mother's X chromosomes the fetus received to predict
whether he will have hemophilia. We do not know the nature or location of
the mutation in the large factor VIII gene. RFLP analysis can help.

To begin we need DNA samples from the fetus, the mother, the mother's
father, and as many other members of the mother's family as possible. We
then look for a polymorphism of the X chromosome, preferably one that is
near the factor VIII gene. Let's call a "normal" X chromosome X-1 and an
X chromosome with the polymorphism X-2. We need to find a polymorphism for which the mother is heterozygous—that is, for which she has
both an X-1 and an X-2. Then we find out whether both affected uncles
had the X-2 chromosome. That information would tell us that the mutation
was on the X-2 (not the X-1) chromosome. Analysis by RFLP analysis will
tell us which one of the two possible X chromosomes (1 or 2) the fetus has
inherited from his mother. If it is X-2, he will have hemophilia; if it is X-1,
he will not.

The RFLP analysis is not as difficult as it sounds. We know of many
polymorphisms from which we can choose. Frequently a suitable polymorphism can be found to use as a marker to help us determine which chromosomes a child inherited from which parent. Recombinant DNA techniques
can link genetic polymorphisms with specific diseases, sometimes long before anything about the actual genetic cause of the disease is known. Having the polymorphic genetic marker does not mean that having the illness is
a certainty. Linkage between a polymorphic genetic marker and a disease
just means that the observable change in the DNA frequently is inherited
along with the illness.

When a genetic polymorphism linked to a disease is discovered, the avenue for further research is clear. Researchers try to find markers that have
tighter and tighter links to the disease. This means trying other polymorphic
markers and finding out if they are better predictors of the disease state. A
tighter link between the marker and the disease means that the two are
less-often separated by genetic recombination at meiosis. This distance can
be measured in centiMorgans. Finding a marker that is closely linked to a
disease is not always possible and never easy. First, the disease state itself
may not be defined precisely. A "specific" diagnosis may actually represent
a mixture of several diseases. Misdiagnosis may add to the confusion. Second, tight genetic linkage may be demonstrated in one large family group
and be completely absent in another.

As researchers find genetic markers that are more tightly linked to a specific diagnosis, there is hope that the specific gene causing the disease will soon be found. When the linkage has been shown to be very tight, the search for a gene is hot. Now researchers start sequencing the genome around the marker. If sequences are found that have the anatomy of a gene, then things are looking good. The next step in research would be to find exactly what mutation occurs in the genome of patients. This mutation must be absent from the genome of unaffected individuals. Usually there is more than one mutation causing a disease. That is, many patients show an A to a T change at a specific site, while other patients show a completely unrelated change hundreds of base pairs away. The genetic alterations are the next clue in finding out what goes wrong in the mutant protein of the patients. The search for the genetic basis of cystic fibrosis demonstrates dramatically the base principles we have just considered.

Cystic Fibrosis

History of the Discovery of the Cystic Fibrosis Gene

The search for the cystic fibrosis (CF) gene is an intriguing story with instances of competition and cooperation. The history of the discovery of the CF gene demonstrates how rapidly scientific progress occurs in molecular medicine. The pace is frenetic, but also exhilarating for those involved. The results are bewildering, since there is so little time between the basic science advances and clinical applications. The story of the search for the CF gene was reported as news in the press ahead of the publication of research results in scientific journals. Parents concerned about cystic fibrosis were calling their doctors even before the medical journals were able to report the facts. The story of CF reflects the extent to which molecular medicine has caught the public's attention.

Cystic fibrosis is a very common inherited disease. Research into the nature of CF has had extensive support from the Cystic Fibrosis Foundation. These circumstances drew attention to CF and made it one of the first inherited diseases to be subject to molecular studies. To begin, pedigrees of families with a history of cystic fibrosis were reviewed. A large panel of polymorphic markers was used to screen available DNA samples from each individual in every pedigree. Soon, these data revealed that the gene was located on chromosome 7. By 1985, more linkage data localized the gene to a region on the long arm of chromosome 7 called 7q3.1. The gene was felt to be within a region spanned by two markers called met and J3.11. Figure 6.4 shows a diagram of chromosome 7 and the progressive steps in the localization of the CF gene.

There was quite a bit of excitement at this time. The region between met and J3.11 is less than 2 million bases (2Mb), which is a genetic distance of less than 2cM. This means two things. First, a genetic distance of 2 cM means that only 2% of the time will genes in this region be separated at meiosis. Genetic counselors could use studies of linkage to met and J3.11 to

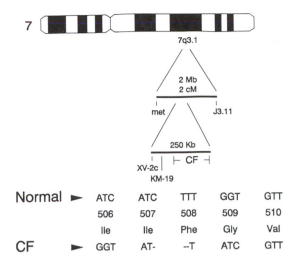

FIGURE 6.4. A diagram of chromosome 7 shows the major genomic error that results in cystic fibrosis, a common autosomal recessive disease.

give a nearly definitive diagnosis of carrier status in families with CF. Second, a physical distance of 2Mb means that researchers were getting close enough to the gene so that luck might play a big part in who actually got there first. It almost did not seem fair. The labs that had spent years finding the general location of the CF gene were now being asked to give their probes and DNA to newcomers who might get most of the credit for doing the last piece of work. Yet holding back on distributing scientific information and tools could certainly not be justified to CF families, who care little about credit, but want a cure. It was like the start of the California Gold Rush. Only in this scientific gold rush, the first prospectors who had done the long work in finding the gold fields were now being asked to give away maps and mining tools to the flood of newcomers. History shows that the scientists involved followed a fairly altruistic course. They ran a fine line between announcing too much too soon, before it had been adequately tested, and being accused of holding too much back. The CF gene was announced as having been discovered several times between 1987 and 1989, but these was false alarms.

Two new markers, XV-2c and KM-19, were found quite close to the CF gene itself, but they were not the gene. Furthermore, a phenomenon called linkage disequilibrium was noted with respect to these new markers. Linkage disequilibrium means that these markers are not distributed randomly on chromosomes throughout the population as one would expect. Disequilibrium suggests that most instances of CF might be due to only a single mutation. Furthermore, the CF mutation may have arisen not too long ago in the time scale of evolution, as only a few recombinant events have occurred between the markers and the disease.

Finally, in 1989, the CF gene was found. Careful studies proved that this time the true gene had been identified. A big part of knowing that the real CF gene had been identified was finding this gene to be mutated in patients with CF. Two laboratories, both of which had been major contributors to the accumulating research, announced in mid-1989 that the mutation causing CF had been found. The mutation was called delta F508. Figure 6.4 shows that exact DNA sequence in the normal genome and in the gene of a CF patient. The mutation is a deletion of three bases from the gene: one base in codon 507 and two bases from codon 508. This causes the deletion of a single amino acid, the phenylalanine residue normally present at the 508th position of the 1,480-amino-acid chain, which is the protein product of the CF gene. The delta F508 mutation, however, was found to be the cause for only approximately 70% of mutations on chromosomes carrying the CF defect.

As important as the discovery of the delta F508 mutation was, researchers subsequently realized that there had to be other mutations that also made the gene nonfunctional and produced the same disease. In a short time, the entire sequence of the CF gene was known. From this nucleotide sequence, a hypothesis regarding the structure of the CF protein could be made. Once the structure of the CF protein was known, it became possible to anticipate where mutational sites would have an adverse effect on the protein's function. Laboratories began to find other mutations besides delta F508 in CF patients. As many as 30% of mutations in the CF gene are other than the delta F508, each of these alternate mutations is comparatively rare, with frequencies less than 1%.

In the space of 4 years, genetic studies in a number of laboratories through individual and cooperative effort have located and mapped the entire cystic fibrosis gene. The mutations that make up most cases of the disease have been found and studies on the function of the protein and how to correct mutated forms are under way. This serves as an example of the power of molecular studies to characterize fully a complex disease and lead to new exact methods for diagnosis and possible treatment.

Genetics of Cystic Fibrosis

Cystic fibrosis is a common autosomal recessive genetic disorder in North American whites affecting approximately one in 2,500 infants. The disease results when a fetus receives two defective copies of the CF gene on chromosome 7. The CF gene codes for a large membrane-bound protein that affects the movement of chloride ions in and out of the cell. In patients with CF this protein is defective, and the secretions of glandular cells are higher in salt content as well as lower in water content, and hence more viscous. The high salt content of sweat is the basis of the sweat chloride test for the diagnosis of CF. The heavy, viscous nature of gland secretions is the major source of pathology in CF, leading to mucus plugs in the lung and in the pancreas. Pneumonia and pancreatitis are the main clinical manifestations of CF. Patients with CF will frequently die in childhood unless good continuous medical care is provided. With antibiotics and hydration as the main therapies, affected individuals can live into early adulthood.

The incidence of CF, one in 2,500, implies a carrier frequency of one in 25 persons among North American whites. Now that the CF gene has been located and some of the mutations responsible for its malfunction have been determined, diagnosis of CF and of carriers is quite feasible using molecular techniques. The gene mutation is present in all cells in the body, so for analysis of carrier status in prospective parents, blood lymphocytes supply an easily obtainable source of DNA. Adults are carriers when they have one normal and one defective gene. In fetal tissues, one is looking for two defective copies of the gene, which would indicate that the fetus will have the disease and not simply be a carrier.

Figure 6.5 shows the results of a PCR-based method for the detection of CF in an affected family. The mother, father, and three children were studied along with fetal tissues obtained by chorionic villus biopsy from a pregnancy 20 weeks in progress. The fetus is denoted by an open diamond symbol. Note that both parents are carriers, a necessary condition for the fetus to have the disease. The oldest child is neither affected nor a carrier, the second child is a carrier, and the youngest child has CF. The diagnosis of CF in their third child is what led this family to seek genetic studies and counseling. The fetus of the current pregnancy was found to be affected. Genetic counseling began before the prenatal testing was instituted and continued after the information that the developing fetus would be affected by CF.

The ability to diagnose CF based on a sample of fetal tissues or to detect carrier status from a blood sample provides an important new test for

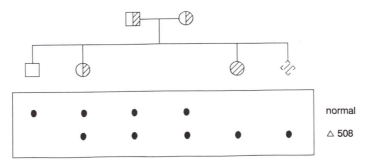

FIGURE 6.5. A pedigree of a family undergoing genetic testing for cystic fibrosis is shown above a dot blot of the PCR products. The presence of a normal and/or a delta F508 PCR product reveals the genetic status that is marked on the pedigree.

this disease. The mutation delta F508, which is detected by a PCR-based technique, occurs in only 70% of CF chromosomes. Some of the other "minor" mutations produce an identical clinical form of CF; other, less-frequent mutations, are associated with very mild forms of CF. If testing for the delta F508 mutation is the only test done for carrier status or for prenatal detection in a fetus, there is a significant possibility that some cases of CF will be missed. Further genetic analysis including linkage studies of RFLP can improve the accuracy of detecting all forms of CF.

The use of molecular diagnostic testing in CF evokes complicated scientific, clinical, and even ethical questions. Since this disease is at the forefront of interest in human clinical genetics, everyone wants to see the "right" thing done to serve as a precedent for future genetic testing. As you might imagine, there is considerable debate as to what is the "right" thing. Should all prospective parents be advised to be screened for carrier status of CF? At the moment, given the significant possibility for false negatives, most geneticists say, "Not yet." If a family already has one affected child with CF, or if a prospective parent comes from a family with a history of CF that indicates possible carrier status, then genetic counseling and possible testing are indicated.

Sickle Cell Anemia

Sickle cell anemia, as discussed earlier in this chapter, results from the inheritance of two mutant alleles for the beta chain of the hemoglobin molecule. Sickle cell trait, which is generally not associated with any clinically abnormality, occurs when an individual is heterozygous, having one mutant and one normal allele. Individuals with sickle cell trait have red cells with a mixture of hemoglobin A and hemoglobin S. Hemoglobin S results from a single nucleotide mutation in the sixth codon of the beta-chain gene as is shown in Figure 6.6. At the 20th base position a T replaces an A, which results in the codon being translated into a valine rather than a glutamine as the beta chain is synthesized. Hemoglobin C, another mutant allele also shown in Figure 6.6, results from an error at the 19th base position in the same codon, resulting in the glutamine residue being replaced with lysine. The change from glutamine to valine for hemoglobin S or to lysine for hemoglobin C results in molecules with different electrophoretic mobilities and abnormal function when compared with the normal hemoglobin A. Hemoglobin S polymerizes at low oxygen, resulting in the characteristic sickled red cell. Sickle cell crisis occurs when these cells block the capillaries in tissues with low oxygen tensions. All of the clinical aspects of sickle cell anemia result from the single point mutation, in codon 6. The pathology for hemoglobin C disease is similar.

Let us assume that a prospective mother and father both of whom have sickle cell trait wish to know if the fetus of a pregnancy in progress will have sickle cell disease. The parents were diagnosed as having the trait in a

codon no.		5	6	7
base no.			20	
hemoglobin	A	CCT	GAG	GAG
		Pro	Glu	Glu
hemoglobin	S	CCT	G**T**G	GAG
		Pro	Val	Glu
hemoglobin	C	CCT	**A**AG	GAG
		Pro	Lys	Glu

FIGURE 6.6. The genetic sequence for a portion of the beta chain of hemoglobin is shown for the normal hemoglobin A allele and for the mutant S and C alleles. Both of these abnormal hemoglobins are the result of a single-base error.

health screening program that utilized a simple solubility test to detect the carrier state. Since the parents both carry the trait, we know that statistically there is a one-in-four chance that the fetus will have sickle cell disease. There is also a two-in-four chance that the fetus is only a carrier, and a remaining one-in-four chance that the fetus is not even a carrier of the hemoglobin S gene. But the parents want more precise information before deciding whether to continue the pregnancy. If we take a small sample of fetal tissue either through amniocentesis or chronic villus biopsy, we can provide that more exact diagnosis.

Southern blot analysis of DNA isolated from the fetal tissue sample provides a simple diagnostic method to find the point mutation that causes sickle cell anemia. It is worth noting that the more simple and standard electrophoresis or solubility procedures used to diagnose sickle cell anemia cannot be applied to this problem of prenatal diagnosis. The fetus has not yet begun to synthesize hemoglobin from the beta-chain locus. During fetal life, a variant form of hemoglobin is made from different genes. Within the first 6 months of life, infants switch their hemoglobin to the adult form, and it is only then that hemoglobin S is noted in patients with the mutant allele. For this reason, prenatal diagnosis of sickle cell anemia requires that we examine the genome directly. Fortunately, as has become apparent in the examples demonstrated in this book, probing the genome is not so hard to do.

Let us reflect for a moment on what is required to detect the beta S mutation. In a DNA sample that represents the entire human genome of a patient, we must determine if the 20th base pair in one gene on chromosome 11 has been switched from an A to a T (as diagrammed in Fig. 6.6). Recall our analogy comparing the human genome to a library in Chapter 1. Our problem is equivalent to finding a single specific spelling error in an entire

| Mst II | C C T N A G G |
| Dde I | C T N A G |

| Hemoglobin A | C C T G A G G A G |
| Hemoglobin S | C C T C **T** G G A G |

FIGURE 6.7. Two restriction enzymes, MstII and DdeI, will cut the genetic sequences in hemoglobin A, but not in hemoglobin S.

library. Again as in Chapter 1, we consider the problem of finding an error in the Preamble to the Constitution by a spell-checking process equivalent to a Southern blot analysis. Now we will employ the method to solve an actual clinical problem. Since we know the genome sequence for the hemoglobin beta chain, we can look for a restriction enzyme whose digestion pattern will be disrupted by the substitution of a T at the 20th base position. Figure 6.7 shows two restriction enzymes that will work nicely. The restriction enzyme MstII recognizes the sequence CCTNAGG, while another restriction enzyme DdeI recognizes CTNAG. The N means that any base in this position is acceptable. Neither of these enzymes will be able to digest the mutated sequence of beta S hemoglobin, where T replaces A. That restriction enzyme site is lost in the hemoglobin S gene. We take samples of the fetal DNA and digest it with *Mst*II. DNA from an unaffected individual and DNA from persons with sickle cell trait and sickle cell anemia are processed in parallel as positive and negative controls. After digesting the DNA, we electrophorese the samples through an agarose gel to separate the fragments according to size. (See Chapter 3 for a detailed description of the Southern blot procedure.) The gel is then blotted to remove the DNA onto a piece of paper. The blot is probed with a radioactively labeled marker of the hemoglobin gene. Figure 6.8 shows a diagram of the autoradiograph of the Southern blot that we obtain. Looking at the control lanes, we see that

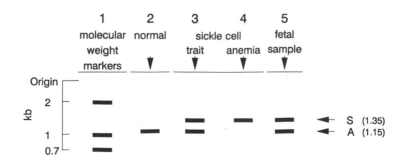

FIGURE 6.8. A Southern blot for the prenatal detection of sickle cell anemia. The DNA has been electrophoresed after digestion with *Mst*II and then blotted and hybridized with a radioactive probe to the hemoglobin gene.

we get one piece of size 1.15 kb for the hemoglobin A gene and one piece of size 1.35 for the hemoglobin S gene. In the sample from a person with sickle cell trait (where both genes are represented), we get both pieces. Lane 5 in Figure 6.8 shows that the sample from the fetus reveals a pattern of sickle cell trait. The fetus is a carrier only. The pregnancy will not produce a child with the severe clinical disease associated with homozygous hemoglobin S.

Issues Surrounding Molecular Prenatal Diagnosis

Having successfully completed a prenatal diagnosis of an inherited disease by using Southern blot analysis, we might be expected to sit back and reflect on the power of modern technology and the advances that may occur in molecular medicine. That is what this book is about. However, this is an equally appropriate time to consider some of the issues — technical, medical, and ethical — that surround molecular prenatal diagnosis.

First, the technical. Are we absolutely sure when we tell these parents that their unborn child will not have sickle cell anemia that there is no possibility for error? No, there are possible sources of error! What if our fetal sample was contaminated with maternal blood? This is a common occurrence. Then our fetal DNA sample would be contaminated with DNA from the mother. We know that the mother has sickle cell trait, and the result that we observed in lane 5 in Figure 6.8 for the fetal sample could be due to the mother's DNA being mixed in! This is an important but not insurmountable technical problem. We can measure how much maternal blood is present in the fetal biopsy sample by looking at any one of several laboratory parameters. Fetal red cells are much larger than adult red cells. So if there is blood in our biopsy, we can determine the source. Alternately we can stain the blood cells in the biopsy for fetal hemoglobin. If they are contaminated cells of maternal origin, they will stain negatively. After we have done these tests, we can determine if our biopsy of the fetal tissue is sufficiently uncontaminated to allow for a Southern blot analysis. Only the white blood cells contain DNA; red blood cells from the mother have no DNA. White cells are 0.1% as frequent in blood as red cells, so the amount of maternal DNA in the sample will be very small if only a small amount of maternal blood is present. If less than 5% of the sample DNA is contaminated with maternal DNA, the Southern blot analysis will not be adversely affected. If there is only a small amount of maternal blood present, we can go ahead. An alternate method of genomic analysis is PCR. The PCR is much more sensitive than Southern blot analysis. For PCR, even 0.1% contaminating maternal DNA might give a false result.

Another source of error in genetic analysis for prenatal diagnosis is much more prosaic but produces false negatives and positives just the same. Sample identification is difficult in DNA analysis. The original biopsy will be handled and aliquotted many times before the final result is seen in lane 5 of Figure 6.8. Strict sample identification must be maintained throughout.

Some early prenatal analyses and some forensic analyses were found to be erroneous due to improper sample identification. If RFLP analysis is used to detect a genetic disease, samples from family members as well as the patient are required. This introduces several other possible sources of error. Family members must be available, and correctly identified. Nonpaternity (the father not being the biological parent) must be considered. Even when multiple family members are analyzed along with the patient, the result may be noninformative. A noninformative analysis means that the markers around the suspected genetic error were not distributed in such a way that inheritance can be predicted. Quality assurance procedures for all genetic studies must be very tight, as these are still difficult procedures to do and the results of an error can be grave.

The medical and ethical issues surrounding this example of prenatal diagnosis of sickle cell anemia are less well defined than the technical issues, but are more important. For any medical diagnostic procedure, the issue of cost-benefit ratio must be raised. How safe is amniocentesis or chorionic villus biopsy? This fetus has only a one-in-four chance of having sickle cell anemia. How much do we put the pregnancy at risk in trying to improve the precision of our diagnosis? If the fetus is found to be affected, do the parents plan to terminate the pregnancy? What is the purpose of the procedure if they do not? Prenatal molecular diagnosis is expensive. The cost of helping this one couple would pay for screening hundreds of persons for sickle cell trait. Is this technology a wise use of finite medical resources?

Finally, we must ask, "Is abortion an appropriate or ethical tool for preventing the birth of genetically disadvantaged infants?" If so, for what diseases? To terminate a pregnancy to prevent the birth of a child with a debilitating and usually fatal illness such as sickle cell anemia or cystic fibrosis might be justifiable to some. But it is not justifiable to others. In Chapter 7 we will consider genetic engineering and a possible means of curing a child with sickle cell anemia, which would make abortion unnecessary. Is prenatal diagnosis with contemplated abortion justifiable in diseases with less morbidity and mortality? What about Down syndrome, dwarfism, or neurofibromatosis, which have much more variable clinical outcomes? Going on from there to the opposite extreme, what about prenatal diagnosis for selecting advantageous traits like intelligence, good looks, or athletic prowess? The answers to these questions are not certain, and even a discussion of these issues is beyond the scope of this book. Nevertheless the issues must be raised. Molecular medicine is revolutionary not only in its ability to improve diagnosis but in the potential impact of this technology on our society.

Familial Hypercholesterolemia

Another genetic disorder, familial hypercholesterolemia, demonstrates how discovering the genetic basis of a disease can greatly increase our understanding of the pathology and guide future treatment. Familial hypercholesterolemia is an autosomal dominant disorder. A pedigree of one family affected with this disease was shown in Figure 6.1. This disorder is relatively common, affecting one in 500 individuals. The principal clinical manifestation of this disorder is accelerated atherosclerotic cardiovascular disease. Among patients who have suffered myocardial infarction before the age of 60, the incidence of familial hypercholesterolemia is found to be one in 20. The disease is characterized by an elevated serum cholesterol in the 300 to 600 mg/dL range and an elevated low-density lipoprotein (LDL), usually greater than 200 mg/dL. Patients with familial hypercholesterolemia frequently have visible xanthomas as a sign of the elevated lipids. Individuals homozygous for familial hypercholesterolemia are rare, approximately one in a million. Homozygotes are more severely affected, frequently suffering manifestations of atherosclerotic cardiovascular disease in childhood and death from this disorder before age 30.

Individuals who are heterozygotes for familial hypercholesterolemia have a defect in forming functional LDL receptors on the surface of their cells. In heterozygotes, there is a decrease in the number of functionally normal receptors; homozygotes have an absence of receptors. When LDLs bind to a functionally normal LDL receptor on the cell, the LDL particle is internalized into the cell by a specialized structure called a coated pit. When the LDL is internalized via one of these coated pits, it becomes degraded to free cholesterol. This free cholesterol is an important intracellular regulatory signal that results in a decreased cellular uptake of exogenous cholesterol. In heterozygotes with abnormally functioning LDL receptors, the signal to limit the uptake of exogenous cholesterol is still received, but only at high levels of LDL. Thus, these individuals hae a qualitatively normal homeostatic mechanism, but if functions only at higher levels of LDL and cholesterol. This results in accelerated atherosclerotic disease.

The genetic defect in familial hypercholesterolemia is not a single mutation in all affected individuals. A very large number of different mutations have been found in numerous portions of the gene. The study of these mutations has been instrumental in gaining a detailed understanding of LDL and cholesterol biosynthesis and cellular utilization. Four main classes of mutations have been found. The most common mutation is a block at the gene level resulting in a failure to synthesize the LDL receptor. In the second class of mutations, the receptor is synthesized, but the protein molecule is not transported from the endoplasmic reticulum to the Golgi region of the cell. The effect is that the receptor never reaches the cell surface. In the third class of mutations, the receptor is normally located on the surface of the cell but fails to bind LDL due to a molecular defect in the structure of the receptor. The fourth class of mutation seen in familial hypercholesterolemia is characterized by the production of a normal LDL receptor that fails to attach in the critical anatomic region of the coated pit on the surface of the cell, thus blocking its function.

The understanding of the LDL and cholesterol biosynthesis pathways and the interaction of the LDL receptor suggests the rational use of various therapies. Dietary restriction decreasing exogenous cholesterol does produce some decrease in serum cholesterol and LDL levels. An alternate therapy is to interfere with the enterohepatic circulation of bile (rich in cholesterol) by using ion-exchange resins such as cholestyramine. It is also possible to interfere with endogenous cholesterol biosynthesis by the drug lovastatin. Taken together, these therapies, individually or combined, can successfully decrease the blood LDL and cholesterol concentrations to a level approaching the normal range in individuals with heterozygous familial hypercholesterolemia. In homozygotes where no functional LDL receptors are present, these therapies are less successful because the pathway is qualitatively broken rather than quantitatively deranged. Homozygotes with familial hypercholesterolemia have been treated with liver transplantation to provide them with a source of cells that contain a functioning gene. This sometimes must be combined with heart transplantation to remove the damage already present to the coronary arteries secondary to high cholesterol levels.

Hypercholesterolemia is of course a very relevant medical condition, not only in the one in 500 individuals with familial hypercholesterolemia but also in the general population. The understanding of the LDL receptor that was obtained in part by studying individuals with hereditary defects can lead to improved treatment for the general population. Some races of humans have been found where the blood LDL cholesterol concentration is at a lower setpoint than is the norm in Western populations. In these individuals the LDL receptor is saturated at lower levels than that seen in other individuals. Further studies of the genetics of the LDL receptor might reveal that a good deal of the variation of serum cholesterol levels is due to polymorphisms in this gene system. Genetic analysis of the LDL receptor gene system might provide prognostic information in childhood, allowing physicians to begin to suggest dietary and drug therapies to prevent the buildup of atherosclerotic cardiovascular disease. Finally, there is the possibility that genetically engineered cells with a hyperfunctional LDL receptor gene could be introduced into the liver or other sites, which would result in an overall lowering of serum cholesterol levels and protection from atherosclerosis. In the vast majority of people, the serum cholesterol level is not elevated due to a "known" mutant allele as in familial hypercholesterolemia. More likely the serum cholesterol level is the result of multiple factors including genetic polymorphisms, diet, and habits. However, the study of cholesterol biosynthesis in the familial form of hypercholesterolemia has greatly clarified our total understanding of atherosclerosis.

Mitochondrial Eve—An Aside

Recombinant DNA technology has so sharpened the precision of genetic analysis that, as a tool, genetics is finding applications in entirely new fields of investigation. The application of molecular genetics to anthropology is, at least to me, quite a surprise and has produced quite an unexpected outcome. As an aside, I would like to relate the controversy of the so-called Mitochondrial Eve. The methods employed in this research, if not the subject matter, will already be familiar to us from the study of the molecular genetics of medicine earlier in this chapter.

The surprising and controversial Mitochondrial Eve theory put forward a few years ago has received much debate, and further experimental data have both supported and refuted the initial conclusion. The Mitochondrial Eve hypothesis states that the degree of genetic similarity between humans (or other species) can be quantitated by comparing the number of mutations within the mitochondrial DNA genome.

The genes of an infant are inherited approximately equally from the father and the mother. The genetic shuffling that occurs at each successive generation provides the fetus with two copies of every gene out of a possible four choices from the parents' genome. This sexual reproduction with genetic recombination increases the diversity of genetic response to evolutionary pressures.

The inheritance of mitochondrial DNA (mtDNA) is an exception to sexual reproduction. Mitochondria, subcellular organelles essential for aerobic metabolism, are thought to have at one time been primitive bacteria that became obligate parasites of eukaryotic cells. Mitochondria have many of the components of bacteria including a limited amount of their own DNA. Of singular importance to the Mitochondrial Eve theory is the realization that mtDNA is inherited only from the mother with no contribution from the father at the time of fertilization. When the sperm penetrates the egg, no mitochondria enter. The zygote's only source of mtDNA is from the mother. A mother passes her mtDNA component onto her daughters and sons. The sons cannot pass this genetic information any further; the information can only be passed on by the daughters. The mtDNA is a circular molecule, only 16,500 nucleotides in length, which codes for ribosomal and transfer RNA as well as some of the electron transport chain enzymes present in the mitochondria. Other enzymes in the mitochondria are encoded by the cell's nuclear DNA.

Investigators at Berkeley used mtDNA studies to trace human ancestry, capitalizing on the little-appreciated fact that mtDNA is passed from generation to generation solely from the mother (Cann et al., 1986). They took mitochondrial DNA samples from 147 different individuals chosen from as many of the anthropologically diverse groups as possible among the earth's population. Australian Aborigines, American Indians, Black Africans, Northern Europeans, and others were sampled. The mtDNA from each individual was subjected to an extensive RFLP mapping. A comparative analysis of mutations in each sample was carried out. The Berkeley investigators formed an evolutionary tree, assuming that the fewer mutational differences between two individuals, the more closely they are related.

The surprising result was that all the branches on this tree join the trunk quickly. Quickly means that using the assumed rate of mutations in human mtDNA, all of the individuals tested could be considered to have a common African maternal parent at a time in the recent evolutionary past. These investigators concluded:

> All these mitochondrial DNAs stem from one woman who is postulated to have lived 200,000 years ago, probably in Africa. All the populations examined except the African population have multiple origins, implying that each area was colonized repeatedly.

The African ancestor was named Mitochondrial Eve by the press. The finding of a convergence on the evolutionary tree to a single common maternal ancestor who lived in sub-Saharan Africa only 200,000 years ago is certainly dramatic.

Many potential flaws in this theory have been extensively debated. The 200,000-year estimate back to the common ancestor could be significantly in error if the assumed mutation rate for mtDNA was wrong. Mitochondrial DNA has a higher mutation rate than does nuclear DNA, because mitochondria lack the extensive DNA repair mechanism that is present in the cell's nucleus. The selection of the 147 individuals might not be sufficient to show other diverse evolutionary trends among humans. In addition, any branch of the evolutionary tree tested using mtDNA is lost whenever a generation occurs in which only sons are born. This leads to many potential dead ends for this type of analysis. Some anthropologists are disturbed that the fossil record of man's evolution does not match the short time span suggested by the DNA record. A great deal of discussion and further data over the last several years have not confirmed or repudiated the Mitochondrial Eve hypothesis. The model has raised important questions about rates of mutation, "conserved" sequences, and genetic drift. Nevertheless, the analysis of mtDNA for the study of evolution in many species is now generally accepted. Recombinant DNA technology has become an accepted and highly quantitative tool for studies in anthropology, evolution, and ecology.

Conclusions

The power of recombinant DNA technology has made possible the direct investigation of human genes. Our appreciation of the genetic basis of many diseases has soared as a result of the new information that has become available. Severe hereditary biochemical defects can now be detected prenatally through analysis of fetal biopsies. Carrier status for life-threatening recessive genetic diseases like CF can be determined in prospective parents. The genetic basis of many common conditions such as hypercholesterolemia can now be appreciated as part of complex multigenic systems. In the next chapter, we consider some instances of genetic engineering and how mutations may be corrected.

Bibliography

Antonarakis SE (1989) Diagnosis of genetic disorders at the DNA level. N Engl J Med 320:153–163.

Budowle B, Baechtel FS, Giusti AM, Monson K (1990) Applying highly polymorphic variable number of tandem repeats loci genetic markers to identity testing. Clin Biochem 23:287–293.

Cann RL, Stoneking M, Wilson AC (1986) Mitochondrial DNA and human evolution. Nature 325:31–36.

Chang JC, Kan YW (1982) A sensitive new prenatal test for sickle cell anemia. N Engl J Med 307:30–32.

Cohen J (1994) Will molecular data set the stage for a synthesis? Science 263:758.

Gelehrter TD, Collins FS (1990) *Principles of Medical Genetics*. Williams and Wilkins, Baltimore.

Highsmith WE, Chong GL, Orr HT, Perry TR, Schaid D, Farber R, Wagner K, Knowles MR, Warwich WJ, Silverman LM, Thibodeau SN (1990) Frequency of the Phe508 mutation and correlation with XV.2c/KM-19 haplotypes in an American population of cystic fibrosis patients: Results of a collaborative study. Clin Chem 36:1741–1746.

Kirby LT (1990) *DNA Fingerprinting*. Stockton Press, NY

Lemna WK, Feldman GL, Kerem B, Fernbach SD, Zevkovith EP, O'Brien WE, Riordan JR, Collins FS, Tsui L, Beaudet AL (1990) Mutation analysis for heterozygote detection and the prenatal diagnosis of cystic fibrosis. N Engl J Med 322: 291–296.

Pauling L, Itano HA, Singer SJ, Well IC (1949) Sickle cell anemia, a molecular disease. Science 110:543–548.

Shibata D (1993) Identification of mismatched fixed specimens with a commercially available kit based on the polymerase chain reaction. Am J Clin Pathol 100:666–670.

Wertz DC, Fanos JH, Reilly PR (1994) Genetic testing for children and adolescents. JAMA 272:875–881.

CHAPTER 7

Immune System and Blood Cells

Introduction

The cell biology of the immune and hematopoietic systems is incredibly rich in detail, involving the generation and interaction of a large number of different cell types. Our understanding of the details is incomplete, yet enough has already been seen to appreciate the wide range of the body's repair and defense mechanisms. We know more about the biology of blood cells than we do about most other tissues in the body because of the ease with which we may sample them. Hematologists have had a major advantage in that the blood is so accessible for microscopic, and now, molecular examination.

The generation of the major classes of cells that make up the immune and hematopoietic system is drawn schematically in Figure 7.1. Each differentiated cell type is derived from a precursor or stem cell that under appropriate stimulation divides. The progeny of the stem cell continue to divide and also to mature, gaining the characteristic functions of the mature blood cell type they are destined to become. Some of the steps in the path from stem cell to mature blood cell are irreversible, involving rearrangement of the genome. Other steps are modulated by growth factors and the outcome may be varied according to the body's immediate need. In addition to the toipotent stem cell, which is the common precursor to all blood cell types, there are committed stem cell pools that can only produce progeny of a more restricted group of cell types. The CFU-GM (colony-forming unit of the granulocyte/monocyte) is one such committed stem cell. As shown in Figure 7.1, the CFU-GM can generate either granulocytes or monocytes. Finally there are fully committed precursor cells, such as the megakaryoblast, which can only produce one cell type, in this case platelets. We can identify most of the cells demonstrated in Figure 7.1 when we look at a sample of the bone marrow under the microscope. The bone marrow is the primary home of the hematopoietic precursor cells. A combination of morphology, cell-surface antigens, and functional tests allows us to identify

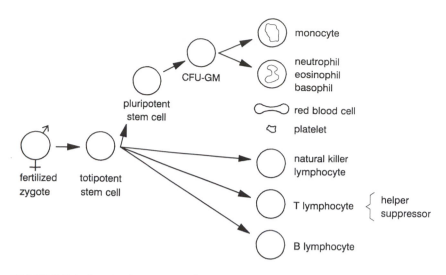

FIGURE 7.1. Stem cells generate all the cell types of the blood and immune systems by a process of proliferation and differentiation in response to growth signals.

most of these precursor cells. The lymphoid cells leave the bone marrow early in their development, passing through the thymus and then on to the lymph nodes. Most of the later steps in differentiation of lymphocytes occur in the lymph nodes.

In this chapter, I discuss the molecular events that help characterize and explain most of the functions of the immune system. We will see, for example, that the diversity of the immune system is derived from rearrangement of the family of genes that encode for the immune proteins. The recent discoveries of the growth factors that influence the activity of the immune and hematopoietic systems is presented. This information is used in the next chapter to help understand the molecular biology of leukemia and lymphoma and new treatments of these diseases. I also discuss genetic engineering of bone-marrow cells as a new avenue of therapy for a number of diseases affecting the immune or hematopoietic systems. The first human gene transplant that used genetically engineered bone-marrow cells is described in some detail. In this landmark event, an attempt to cure an inherited immunodeficiency syndrome was made by replacing a defective gene in the cells of the bone marrow.

An understanding of the molecular biology of blood cells leads the way in demonstrating the clinical impact of recombinant DNA technology on medicine. The blood cells were among the first tissues examined by the microscope; they were the first to be examined by molecular probes, and they are the first to be genetically engineered to reverse human diseases.

Molecular Biology of the Immune System

The immune system has a very complex set of tasks to carry out, and these functions must be accomplished everywhere in the body. We must envision a *cellular ecology* of the immune system to emphasize and appreciate the complex interactions between cell types and growth factors. Figures 7.2 and 7.3 present a more detailed view of the generation of lymphocytes than was presented in Figure 7.1. These figures show that under the appropriate stimulation a lymphoid stem cell can differentiate along either the B- or T-lymphocyte maturation pathways. The surface antibodies by which we recognize the various B and T lymphocytes are shown in Figures 7.2 and 7.3, as they appear with progressive maturation of the cells. These antibodies are labeled with the letters CD, for *cluster designation*, followed by a number. The CD nomenclature is complex, but it reflects the functional complexity of the immune system.

B and T lymphocytes are derived from stem cell precursors in the marrow. Pre B lymphocytes move to the lymph nodes and other lymphoid tissues and mature there. However, pre T lymphocytes must first pass through the thymus, during fetal development. In the thymus, pre T cells are "schooled" in proper function before going out into the blood or lymph nodes. Each individual lymphocyte produces an antibody of unique specificity. Mature B or T lymphocytes undergo clonal expansion, making many more copies of themselves when appropriately stimulated by an antigen or by interaction with another cell. The members of a clone of lymphocytes are all the progeny of a single precursor. The clone produces the same unique antibody as the precursor. All of these interactions, and more that

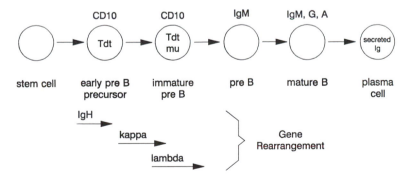

FIGURE 7.2. The differentiation of B lymphocytes from the stem cell shows the progressive steps in gene rearrangement beginning with the immunoglobulin heavy chain (IgH), followed by kappa and lambda light chains. The maturation of the immunophenotype of the cells beginning with the expression of the enzyme Tdt and continuing with expression of CD10, cytoplasmic, and surface immunoglobulins is shown above the cells.

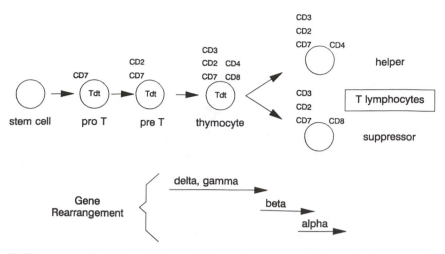

FIGURE 7.3. The differentiation of T lymphocytes also begins with the stem cell. The rearrangement of the genes of the T-cell-receptor gene begins with the delta and gamma chains and then proceeds with beta and alpha chains. The immunopheno-type, which helps identify degree of differentiation, is also shown. Tdt is expressed first, followed by pan T markers CD7 and 2. Later, the markers of the functional classes for helper and suppressor T cells appear (CD4 and CD8).

have not yet been discovered, serve to maintain the immune system's ability to repair cell damage and react to foreign material.

Lymphocytes are capable of reacting to millions of potential antigens. The number of possible antibodies produced by lymphocytes is so great that the human genome does not have enough DNA to encode for all the possible molecules. For years, people wondered how the diversity of the immune response was generated. One theory was that the antibody mole-cule "folded" around an antigen and changed its structure in response to the antigen. We now know that this theory is wrong. Gene rearrangement, as we will see, is the basis for the diversity within the immune system.

Mature differentiated lymphocytes can only generate copies of them-selves. To generate a new clone with a different immune specificity, it is necessary to start over with a stem cell and progress through the successive stages of development. The events that control lymphopoiesis are complex. Some antigens stimulate proliferation directly. However, most antigenic stimulation requires an interaction between at least two cells of the immune system such as a T helper lymphocyte and a mature B lymphocyte for a clone to be stimulated to proliferation and increased antibody production. Monocytes can also process antigens and then influence the immune re-sponse by secretion of interleukins that function as growth factors.

Immunoglobulin Gene Rearrangement

Lymphocytes irreversibly rearrange their genome as they differentiate from stem cell to mature B or T cells. This phenomenon of immune gene re-

arrangement is the basis for the generation of diversity of antibody production within the immune system. Let us consider in detail how gene rearrangement occurs as B lymphocytes mature, in order to produce a
complete immunoglobulin molecule. An immunoglobulin molecule (IgG
type) is composed of two heavy chains and two light chains. Each part of
the heavy and light chains has a number of alternate possible genes that
must be brought together to make a single complete protein molecule. The
genes that encode for the heavy chain are located on chromosome 14; those
for the light chains are on chromosomes 2 and 22. It is necessary to select
one gene from each of the possible alternate genes (which as a group constitute a gene family) and join them together to make a complete gene for the
protein. An analogy is to think of the immunoglobulin gene family as a
deck of playing cards. Each mature lymphocyte constitutes one play of the
hand where five or six cards are drawn from the deck and are laid out as
the genotype for this cell. In this way, the immunoglobulin gene family
consisting of tens of individual genes can be arranged into millions of
possible genotypes, each capable of making a different antibody molecule.

The process of gene rearrangement for the heavy chain is shown schematically in Figure 7.4. The first step in gene rearrangement for the heavy chain
is the combining of the diversity region D gene with a joining region J. The
next step is to fuse this DJ combination with one of the variable portion
genes for the heavy chain, Vh, giving a complete VDJ sequence for the
heavy chain. The rearranged genetic sequences serve as a template for RNA
transcription, which, after splicing, generate an mRNA blueprint for the
immunoglobulin-chain protein. Alternate splicing sites in the RNA can gen-

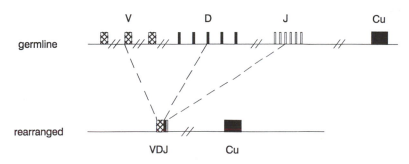

FIGURE 7.4. The rearrangement of the immunoglobulin gene is shown to demonstrate the joining of individual elements from each component class of the gene
family. The process starts with joining of a D and a J element. The intervening
DNA is spliced out and is irreversibly lost. Next a V element joins to the DJ piece.
This VDJ rearranged genome will serve as the DNA template for a portion of the
specific immunoglobulin molecule made uniquely by the lymphocyte bearing this
particular rearrangement.

erate more than one immunoglobulin protein from the same genetic sequence, adding additional genetic diversity.

The immunoglobulin light-chain genes also undergo rearrangement similar to that of the heavy chain. Finally, two heavy and light chains combine into a unique antibody molecule. As the cell matures and rearranges it genome, it discards the other genes that it did not use and thus irreversibly fixes itself as a cell that can now only produce a single antibody. This process is rather complex, but the complexity seems to have a purpose. The immune system is very rich in diversity and there are many steps in the system that can be regulated. Immune gene rearrangement is, however, inefficient. Only one out of three B lymphocytes that attempts to mature successfully recombines its heavy- and light-chain genes into a configuration that can produce an antibody molecule. Part of the reason for this low success rate is that it is necessary to link the genes in phase so that the recombined genetic sequences will constitute an open reading frame for translation of mRNA into a protein.

As far as we know, lymphocytes are the only cells in the body that physiologically rearrange their genes. No other cell has yet been discovered where genes move around as part of a normal process. Lymphocytes rearrange their genes to achieve antigenic diversity. From only tens of genes, millions of antibodies can be produced. This gene rearrangement is so clever that I believe there will be other systems in the body where gene rearrangement will be discovered. For example, rearrangement of cell-surface proteins on neurons of the central nervous system might be a mechanism leading to the extreme complexity of interneural connections.

T-Cell-Receptor-Gene Rearrangement

The molecular biology of T lymphocytes is generally quite similar to B lymphocytes with the exception that the protein produced by gene rearrangement for T cells must stay fixed to the surface as a receptor molecule. Furthermore, the T cell receptor (TcR) acts only indirectly with foreign antigen rather than with the direct binding that is seen in B lymphocytes. The T cell receptors that control immune interaction contain four major protein chains: alpha, beta, gamma, and delta. These four proteins are combined to generate the very complex surface receptor molecules. These T-cell-receptor chains are synthesized from rearranged genomes in a process quite analogous to the rearrangement of immunoglobulin genes in B lymphocytes.

Southern Blot Analysis of Gene Rearrangement

Immune gene rearrangement can be analyzed by Southern blotting. A Southern blot of peripheral blood lymphocytes will show a germline pattern for both the immunoglobulin and the T-cell-receptor genes. The germline pattern reflects the nonrearranged DNA from the unaffected chromosome. Gene rearrangement occurs only on one of the two chromosomes that carry the immune genes. This germline will be overlaid with an invisible smear of multiple different rearrangements. Blood lymphocytes are normally polyclonal, consisting of a mixture of B and T lymphocytes that are primed to react to a wide variety of antigens. Each different lymphocyte gives a different banding pattern from its rearranged chromosome. These polyclonal bands are too faint to be seen on an autoradiograph. However, if a clone of identical lymphocytes is present in the sample, the exact copy of the gene rearrangement present in each cell of the clone will be detected in a Southern blot. The clone will be visualized as an additional band on a Southern blot. Figure 7.5 demonstrates a Southern blot analysis for rearrangement of the heavy chain of the immunoglobulin gene. A probe to the joining region, Jh, has been hybridized to three different restriction digests. In the lanes marked as negative controls (−), a single band representing the germline pattern of DNA from the unaffected chromosome is seen. In the lanes marked as positive controls (+), DNA from a B cell lymphoma was analyzed. Note that in addition to the single germline band, there are one or more additional rearranged bands in each of the three digests with BamHI, EcoRI, and HindIII enzymes. The rearranged bands come from the clonal expansion of a malignant B cell, which is present as the dominant population in the lymph node that was sampled. The lanes marked 1, 2, and 3 are a test of a biopsy from a patient who has not yet been diagnosed. Note that the patient samples show rearranged bands in each of the three digests. This indicates the presence of a clonal population of B lymphocytes consistent with a lymphoma.

Southern blot analysis of immune gene rearrangement can be further appreciated by thinking of the close analogy with the detection of monoclonality in serum protein electrophoresis. The immunoglobulins normally present in the serum are a mixture of several types consisting of IgG, IgM, IgA, and so on. Each type is represented by many different molecules, each coming from a different plasma cell. A serum protein electrophoresis shows only a smear of proteins in the immunoglobulin region. If a population of monoclonal cells is present, each of the cells will produce an identical immunoglobulin molecule and the electrophoretic smear will resolve to have a focused band. Southern blot analysis permits detection of a monoclonal population of B or T lymphocytes by an exactly analogous process. Most lymphomas do not result in the secretion of an immunoglobulin protein and thus would be undetectable by serum protein electrophoresis. Fortunately, we can now detect them by DNA probes. (See Chapter 8 for more information on lymphomas.)

Hematopoiesis

The production of all nonlymphoid blood cells is carried out in the bone marrow. A small number of hematopoietic stem cells, estimated to be as few as 1 million in number or 1 μg of tissue, are responsible for generating

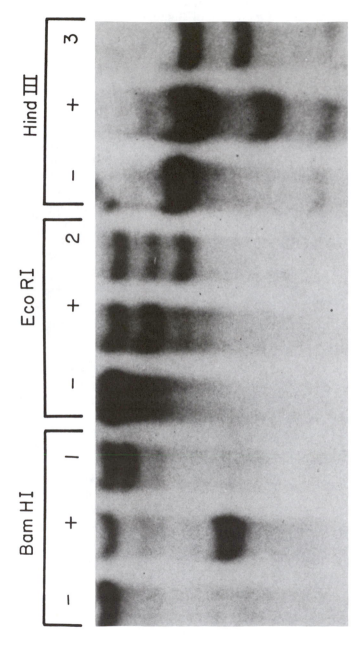

FIGURE 7.5. Southern blot analysis of lymph node tissue for the presence of a monoclonal population of B lymphocytes.

all of the blood cells of the body. The initial stages include between seven and ten cell divisions, leading to a large amplification in the number of marrow cells. During this early phase, the stem cells become committed to a specific lineage. A hematopoietic stem cell that initially is able to produce any kind of blood cell, upon receiving the appropriate signal becomes a committed stem cell capable of producing just one cell type, as is demonstrated in Figure 7.1. There are separate committed stem cells for red blood cells (also called erythrocytes), platelets, and granulocytes/macrophages. The committed stem cell precursors are named CFU-E, CFU-Meg, and CFU-GM, respectively. CFU is an abbreviation for colony-forming unit, which is the experimental assay for counting stem cells.

After the first seven to ten cell divisions, the committed blood cell precursors continue to divide and also begin to mature. Each stem cell now produces two daughters that are more mature and have more of the final characteristics of peripheral blood cells. For example, a myeloblast divides to form two promyelocytes that contain primary granules. Each of the two promyelocytes divides, forming a total of four myelocytes. These myelocytes now have both primary and secondary neutrophilic granules. A few more cell divisions result in the mature neutrophil, ready for release from the bone marrow into the peripheral blood. The same process occurs in the red cell lineage. A pronormoblast divides to form two basophilic normoblasts that divide into four polychromatophilic normoblasts. These normoblasts begin to synthesize hemoglobin in the cytoplasm. The nucleus condenses and is expelled from the cell as the cytoplasm continues to fill up with hemoglobin. The final product after expulsion of the red cell nucleus is the young red blood cell (reticulocyte) that is released from the marrow to the bloodstream. Megakaryopoiesis is similar except that as the megakaryocyte matures, it does not divide into daughter cells. Instead, the cytoplasm of the giant megakaryocyte fragments into large numbers of individual platelets.

Growth Factors and Blood Cells

The process of hematopoiesis is under control of a large number of growth factors. Examples of growth factors include erythropoietin (EPO), interleukins (IL), interferons, and colony-stimulating factors (CSFs). Some factors, particularly interleukins and interferons, are also called lymphokines and monokines. The functions of these molecules overlap and the terminology can be confusing. Growth factors bind to specific receptors on the cell surface. Many of the genes that encode for both the growth factors themselves and for their corresponding receptors have been discovered and cloned. The success in cloning these genes means that recombinant growth factors can be produced in pharmaceutical quantities. The availability of recombinant growth factors such as erythropoietin (EPO) and white-cell-stimulatory factors (GM-CSF and G-CSF) makes possible the specific stimulation of blood cell production. The EPO is used to treat the chronic anemia of renal failure, sickle cell disease, and AIDS. The G-CSF and GM-CSF are used to treat the severe neutropenia associated with chemotherapy for cancer or AIDS-related suppression of bone-marrow function.

The pathways of cell proliferation and differentiation as mediated by growth factors are complex. Many factors are expressed after processing by several intermediate cell types. The growth factors are also codependent on each other, one factor being the stimulus or suppressor of several other factors. For example, tumor necrosis factor (TNF) and IL-1 are both secreted by monocytes in response to certain stimuli such as bacterial endotoxin. Tumor necrosis factor and IL-1 cause T lymphocytes, endothelial cells, and fibroblasts to secrete several CSFs, which in turn affect proliferation of neutrophil, monocyte, and megakaryocyte precursors. The increase in CSFs results in myeloid hyperplasia in the marrow. There is negative feedback as well as positive in this system. The effects of growth factors are modulated by altered levels of protein (in response to switching genes on or off) as well as in response to up- or down-regulation of the number of receptors. The system is complex but contains the possibility for much variation on the part of the immune and hematopoietic systems in response to stimuli.

In the lymphoid system, the interleukins help control not only cell number but cell function. For example, IL-2 predominantly stimulates T cell proliferation, but it also affects B cell and natural killer (NK) cell activity. Monoclonal antibodies to the IL-2 receptor (called CD25 or Tac) bind to T cells, B cells, and macrophages. Soluble IL-2 receptor complexes can also be detected in plasma and are indicative of changing immune function. As antoher example, IL-4, which is secreted by T cells, increases the expression of HLA-DR antigens on B cells as well as increasing the number of activated B cells. The interleukins also help bridge the effector cells of the lymphoid and hematopoietic system. Secreted by lymphocytes, IL-5 is a strong stimulus for eosinophil proliferation. An understanding of the interlinking messages exchanged between cells makes it possible to appreciate the broad diversity of inflammatory cell infiltrates in differing pathologic situations.

The administration of recombinant CSFs dramatically affects the immune response. Our morphologic interpretation of blood, marrow, and lymph node biopsies is also greatly altered by these agents. Four CSFs are now available as recombinant proteins in quantities sufficient for experimental and therapeutic use. These include granulocyte (G-CSF), macrophage (M-CSF), granulocyte plus macrophage (GM-CSF), and multipotential (multi-CSF, also called IL-3) colony-stimulating factors. G-CSF and GM-CSF are administered for the treatment of neutropenia following intensive chemotherapy for cancer. Patients with severe neutropenia following chemotherapy for small-cell carcinoma of the lung, when treated with GM-CSF, rebound with an elevated white blood cell count (20 to 40 $\times$ 10^9/L). Bone-marrow examination shows markedly increased cellularity, approaching 100%, with a myeloid predominance. The recombinants G- and GM-CSF have also been used to treat neutropenia in AIDS patients resulting from zidovudine (AZT) toxicity.

Other clinical trials using recombinant CSFs are under way for the treatment of the pancytopenia seen in certain myelodysplastic syndromes. A number of myelodysplastic syndromes and treatment-related leukemias are associated with alteration of the long arm of chromosome 5. It is intriguing that the genes for IL-3, GM-CSF, M-CSF, and the M-CSF receptor (also called the c-*fms* oncogene) are clustered on the long arm of chromosome 5. Deletion of this portion of chromosome 5 occurs commonly in myelodysplastic syndromes and in treatment-related secondary leukemias.

Therapy of lymphoma with recombinant IL-2 (and sometimes, in addition, with infused activated lymphocytes) is undergoing clinical trials. The IL-2 may force differentiation of lymphoma cells to an extent that shuts off their proliferative capacity. The optimal therapeutic use of IL-2 and other interleukins will require improved understanding of the regulatory pathways still intact in lymphoma cells. In cancer patients IL-2 may also selectively stimulate the immune system. Immune stimulation is especially desirable for the NK function of lymphocytes directed against tumors. This approach is part of a broad new experimental trend in cancer treatment called biological response modifier therapy.

Erythropoietin is the most widely used pharmaceutical agent among the growth factors. Some would consider EPO a hormone, since it is secreted by an organ (the kidney) distal to the target cell (bone-marrow erythroid stem cell). Recombinant erythopoietin is now generally available and is used to treat the anemia of chronic renal failure. Erythropoietin is also under investigation for treatment of AIDS-associated anemia and for the treatment of sickle cell anemia.

Bone-Marrow Transplantation

The transplantation of bone marrow allows for reconstituting the blood cells of the host. Bone marrow is unique among organ transplants in a number of ways. Surgery is not required. Bone-marrow cells are harvested by aspiration from the donor and transplanted by intravenous infusion into the host in a manner identical to blood transfusion. The donor bone-marrow stem cells that are transfused into the peripheral blood "home in" to the bone marrow, probably in response to local factors in the microenvironment within the marrow space of bones. The donor is only temporarily depleted of bone-marrow tissue; regrowth is rapid. A person can in principle be a donor a number of times. The bone marrow consists of many cell types, but its function can be completely regenerated by infusion of a small number of hematopoietic stem cells that are capable of dividing to increase their numbers. Under appropriate growth-factor signals, bone-marrow cells differentiate to produce all the red and white blood cell elements of the body. The bone marrow is an immunologically active organ in that it contains and regenerates lymphoid cells as well as granulocytes, monocytes, red cells, and platelets. Bone-marrow transplantation offers the unique problem of graft-vs.-host reaction because the donor bone marrow contains immunologically active cells that can fight back against the host tissues. This means that in addition to the standard problems of the donor rejecting the organ, bone-marrow transplantation introduces the problem of the organ rejecting the donor. Graft-vs.-host disease can be fatal.

Bone-marrow transplantation is used to treat a number of clinical conditions including (1) hematologic malignancies in which the malignant marrow is killed by intensive chemotherapy and a transplant is given, (2) serious congenital defects of hemoglobin synthesis such as thalassemia major and sickle cell disease, (3) congenital severe combined immune deficiency (SCID) in which bone-marrow transplantation reconstitutes the immune system, (4) metabolic diseases such as Gaucher disease in which improving bone-marrow phagocyte function can diminish the systemic manifestations of the illness, and (5) metastatic cancer in which the bone marrow is killed by the toxic antitumor chemotherapy.

Recombinant DNA techniques have made possible diagnostic tests that aid in bone-marrow transplantation. Using DNA fingerprinting, it is possible to assay a blood sample and determine whether the cells are of host or donor origin. This test can be used to determine whether the engraftment has "taken." Sometimes after bone-marrow transplantation, the marrow becomes chimeric, consisting of both host and donor stem cells. When bone-marrow transplantation is used for the treatment of leukemia, DNA probes can monitor for the presence of a small amount of residual leukemic cells as well as detect engraftment from the donor cells. If relapse occurs, it is possible to tell whether the relapse is the original leukemia or a new leukemia that may have originated in the donor bone-marrow cells due to the effects of chemotherapy. In addition to DNA probes for the monitoring of engraftment, molecular techniques are used to follow growth factor levels.

Bone-marrow transplantation can be either allogeneic (from a human leukocyte antigen [HLA]-matched donor, usually a sibling) or autologous (marrow taken from the patient, stored and treated outside the body, and then reinfused into the patient). Autologous transplants are useful in permitting the intensive chemotherapy necessary to treat metastatic cancer. The prolonged neutropenia following treatment of lung cancer and breast cancer by multidrug chemotherapy has significantly improved with the use of autologous bone-marrow transplantation. Bone-marrow cells are harvested, kept outside the body during the infusion of chemotherapeutic drugs, and then reinfused. Cell separation techniques make it possible to remove small numbers of tumor cells that might be harbored in the bone marrow as microscopic foci of metastatic disease.

Genetic Engineering of Bone-Marrow Cells

The bone marrow is an enticing target for attempts to introduce genetically engineered cells into the body because it is such a large factory for cell production. Genetic engineering of the bone marrow takes advantage of the relative ease of removing and reimplanting the cells. Bone marrow is also ideally suited to genetic engineering, since only a few bone-marrow stem cells are capable of repopulating the entire bone marrow. In experiments with mice, it has been possible to reinfuse a single cell to repopulate the entire bone marrow of a mouse that has had its own bone marrow killed by radiation treatment. In humans, reinfusion of 10^{10} marrow cells, approximately 10 g of tissue, has resulted in repopulating the entire bone marrow, which is close to 1 to 2 kg of tissue in the average adult. It is theoretically possible that the entire human bone marrow could be reconstituted by engraftment of a single stem cell.

Thus, if genetic engineering is carried out on bone-marrow stem cells, not only is it easy to reinfuse the engineered cells that independently seek out their correct location, traveling from the bloodstream to the bone marrow; in addition only a few cells can reconstitute the entire organ and thus greatly amplify the effect.

Sickle Cell Anemia

If a hematopoietic stem cell with a genetic error from a patient with sickle cell disease could have the corrected hemoglobin gene inserted, then reinfusion of the stem cell would result in the patient making hemoglobin A in addition to hemoglobin S, thus curing sickle cell anemia. Cure of sickle cell anemia by genetic engineering offers an alternative to the prenatal diagnosis and possible abortion, as was presented in the chapter 6. The treatment of congenital hemoglobinopathies such as sickle cell anemia or thalassemia by genetic engineering is particularly attractive for several reasons. First, in the case of sickle cell disease, even a partial success in replacing hemoglobin S with normal adult hemoglobin A would likely cure the patient. A mixture of 10% hemoglobin A plus 90% hemoglobin S greatly decreases the chances of sickle cell crisis. It would not be necessary to completely replace the bone marrow with genetically engineered cells; a chimera of the patient's bone cells plus the patient's bone cells that have been genetically engineered would produce a cure.

The hemoglobin A gene has been cloned and its genetic sequences are available for inserting into a target cell. The technical problems include not only inserting the gene in an intact state containing its full sequences, but in addition, placing the gene in a location so that it will be expressed. The gene must be placed into hematopoietic stem cells that when stimulated to produce red cells will result in the inserted gene being turned on to make hemoglobin A. Just inserting the gene in any random place in the genome will not result in its correct expression. The options for the genetic engineer are either to insert the gene in its correct location or to insert the gene with promoter elements ahead of it, which would recognize the correct signals for turning the gene on. Fortunately, the genetic engineer has a great deal of flexibility in correcting hemoglobinopathies. The inappropriate expression of hemoglobin A, if the gene is turned on too soon or too late in erythroid development, is not likely to damage a cell. The successful cure of sickle cell disease by reinfusing genetically engineered bone marrow has not yet occurred, although experimental trials in patients are planned. As we will see in the next section, the first human gene transplant, which has already occurred, was targeted at another hereditary defect. Genetically engineered bone marrow cells were the vehicle for carrying the new gene into the patient's body.

The First Human Gene Transplant

The first human gene transplant was a momentous event in the history of science and medicine. It was carried out quietly and anonymously at the National Institute of Health on September 14, 1990. For the first time humans were not only examining their genetic makeup but were attempting to rewrite it. Genes for all organisms are in a constant state of evolution. Over great time spans, the genetic makeup of a species changes to adapt better to its environment. In 1990, only 15 years after the start of the recombinant DNA revolution, the first gene transplant was made into a patient in an attempt to alter the genetic makeup of an individual to correct an error. The ability of humans to read and reconstruct their own genetic makeup is unprecedented in the scheme of biological adaptation to the environment.

In Chapter 1, I used an analogy comparing the size of the human genome to a large library. One of the flaws in this analogy is that the human genome is not really a library in the sense that it can be read, but nothing can be written into it. With the ability to change the human genome by genetic engineering, the human genome really does become a library. Mankind now may read what is written there, as well as have the power to add new information and to make corrections. The power to alter the human genome is certainly not without danger. We have really only begun looking into the human genome within the last 5 years and already we have seized the courage to consider changing it.

The first human gene transplant was done for purely medical reasons to help cure a child with a lethal disease and allow her a normal life. The events leading up to this human gene transplant were dramatic. The regulatory steps and review of the process by government agencies were technically more complex and took more time than did the genetic discoveries and manipulation of the gene.

The disease affecting the patient was adenosine deaminase (ADA) deficiency—a genetic, inherited disease that creates a clinical state of severe combined immunodeficiency (SCID). It is the cause of approximately 25% of all cases of SCID. Patients with ADA begin to develop recurrent and severe opportunistic infections beginning in infancy. These infections, if untreated, are usually fatal within the first year of life. Even with intensive treatment of each subsequent infection, the condition is lethal. In the past, attempts have been made to place children with ADA into a sterile environment where they never come in contact with the outside world and its myriad infectious organisms.

Adenosine deaminase deficiency is inherited as an autosomal recessive condition. Fortunately it is quite rare, with an estimated incidence of ten per million births. The cause of the immune deficit in ADA deficiency is selective toxicity to lymphocytes, both T and B cell lineages, due to accumulation of metabolites of deoxyadenosine. Lymphocytes are hypersensitive to deoxyadenosine metabolites, especially deoxyadenosine triphosphate (dATP). Lymphocytes are the only cells severely damaged by ADA deficiency syndrome. Normal levels of ADA break down these metabolites and allow the cyclic AMP metabolism of the cell to proceed in a normal fashion. There are several reasons for choosing to correct ADA deficiency by human gene transplant therapy. Drs. Anderson, Blaese, Rosenberg, and their collaborators—the investigators at the National Institutes of Health— wished to do a gene transplant that would not require altering the germline genetic makeup of an individual. That is, they did not want to correct the defect by making the person a transgenic organism. This is for reasons of safety and concern over altering the genetic makeup of a human. In theory ADA deficiency could be corrected by inserting the gene into bone-marrow stem cells only. These stem cells would then supply enough of the enzyme to overcome the deficiency in lymphocytes.

In the past, a few patients with ADA deficiency have been cured by a successful allogeneic transplant from an HLA-matched bone marrow donor. Regrettably, only a small minority of patients are able to be matched with a bone-marrow donor, which limits this therapy. Other reasons for selecting ADA deficiency for a human gene transplant are important tactical reasons. The human ADA gene had been cloned. A very simple defect in the gene results in the loss of enzyme activity. The ADA gene system is not part of a complex multigene process. The transplantation of genetically altered bone-marrow cells transfected with a normal gene for ADA should provide a source of enzyme to newly generated lymphocytes. The overall level of expression of the transplanted ADA gene is not critical since a wide range of levels of the enzyme corrects the defect without introducing any new problems of its own.

To begin the transplant, bone-marrow cells were removed from the patient and taken to the laboratory. There the blood cells were transfected with a correct copy of the ADA gene in a viral vector. One theoretical danger was that the viral vector, when the cells were reintroduced into the patient, might continue infecting other cells. This could cause the ADA transgene, made in the laboratory, to enter other cells besides the intended target. However, animal studies showed that there was no risk of unintended spread of the gene by the vector. After verifying that the patient's blood cells had taken up the gene in the laboratory, the actual transplant itself was undertaken. No news coverage was permitted. The genetically altered blood cells were simply infused back into the patient through an arm vein.

In the ensuing 5 years since the first human gene transplant, the results continue to be good. The first patient continues to have an augmented immune system. Other patients with ADA have also been treated. Most recently the gene transplant is carried out using an improved technique where stem cells from the umbilical cord are saved at birth. These cord blood stem cells are then transfected with the retrovirus that carries the ADA gene. Because these stems cells are immature, they are readily attacked by the retrovirus, and once transfected they continue to divide for long periods of time. Children treated in this fashion, are not expected to need retreatment. The first patients treated with transfected peripheral blood lymphocytes eventually cleared those cells from their blood and needed to be reinfused with new lymphocytes. Patients treated with cord blood stem cells may never require repeat infusions.

Future Directions

Much has been recently discovered about the molecular biology of blood cells, but when will this result in better therapy for immune and blood disorders? There is a sense of an impending breakthrough in therapeutics, which everyone interested in biotechnology is watching most closely. Molecular medicine will likely have its first dramatic applications in the area of hematology. The ability to clone the genes for many of the growth factors and growth factor receptors and the ability to engineer new genes into bone-marrow cells are the tools for these future applications. One can imagine correcting many of the immune defects and hemoglobinopathies by genetic engineering. Other gene defects, including metabolic disorders, may also be amenable to genetic engineering of the bone marrow.

Cancer therapy may also benefit from genetic engineering. Rosenberg et al. (1990) have genetically altered the lymphocytes from a patient with malignant melanoma to make them more immune reactive against the patient's own tumor cancer cells.

Other even more exotic ideas are being considered for genetic engineering of the bone marrow. Taking bone marrow cells and genetically altering them to remove antigenic determinants could make possible the growth of blood cells in culture suitable as "universal donor" material. The transfer of blood cells and bone marrow grown in the laboratory as a therapy for many diseases would be the start of artificial biological organs.

Bibliography

Groopman JE, Molina JM, Scadden DT (1989) Hematopoietic growth factors: biology and clinical applications. N Engl J Med 321:1449–1459.

Kan YW (1992) Development of DNA analysis for human diseases. Sickle cell anemia and thalassemia as a paradigm. JAMA 267:1532–1536.

Kohn DB, Anderson WF, Blaese RM (1989) Gene therapy for genetic diseases. Cancer Invest 72:179–192.

Rosenberg SA, Aebersold P, Cornetta K, Kasid A, Morgan RRA, Moen R, Karson EM, Lotze MT, Yang JC, Topalian SL, Merino MJ, Culver K, Miller AD, Blaese RM, Anderson WF (1990) Gene transfer into humans: immunotherapy of patients with advanced melanoma using tumor infiltrating lymphocytes modified by retroviral gene transduction. N Engl J Med 323:570–578.

Stamatoyannopoulos G, Nienhuis AW, Leder P, Majerus PW (eds) (1994) *The Molecular Basis of Blood Disease*, 2nd ed. WB Saunders, Philadelphia.

Zucker-Franklin D, Grossi CE, Marmont AM (eds) (1988) *Atlas of Blood Cells: Function and Pathology*. Lea and Febiger, Philadelphia.

CHAPTER 8

Cancer

Introduction

Cancer is a disease of uncontrolled cell growth. Cancer cells grow abnormally because of damage to their DNA. This DNA damage garbles the genetic signals for normal growth. The molecular approach to cancer gets to the heart of the matter. Molecular studies have uncovered many of the details whereby a normal cell becomes cancerous. These discoveries have revealed cancer to be a multistep process, involving the progressive loss of control by the malignant cell, failure of DNA repair, and of loss of backup systems for preventing abnormal cell growth. We have discovered oncogenes, which promote tumor growth, as well as anti-oncogenes, which suppress tumors.

This chapter on cancer begins with a description of carcinogenesis at the DNA level. That knowledge is applied to the clinical problem of colon cancer. The molecular steps that lead to the development and progression of colon cancer are considered in detail. Other important clinical examples, including leukemia and lymphoma, where the molecular biology of the disease is well understood, are examined. Molecular diagnostics makes all of these important basic-science discoveries accessible to clinical medicine with significant therapeutic consequences.

Carcinogenesis

Mechanisms

Cancer results from an acquired defect in the DNA of a cell that causes deregulation of the cell's growth processes. The damaged cell transforms from benign to malignant and becomes independent of normal regulatory signals. This transformed cell multiplies into a clone of malignant cells, eventually developing into a tumor. The malignant tumor infiltrates adjacent tissues and metastasizes. There are several mechanisms by which the DNA of a normal cell becomes damaged, resulting in malignancy. The process is called carcinogenesis, and the damage may occur as a single event or as an accumulation of multiple events. Carcinogenesis can result from (1) a point mutation in DNA, (2) gene deletion, or (3) chromosomal translocation resulting in gene rearrangement. The events that cause each of these errors are, in part, understood.

Point Mutation

Point mutation, meaning the replacement of a single correct nucleotide within a gene by an incorrect one, can occur as a result of one of several mutagenic events. Mutagenic chemicals enter the cell and bind to DNA, forming a structure called a DNA adduct. This adduct will interfere with the correct copying of DNA the next time the cell divides. After the cell divides one daughter cell will have an incorrect nucleotide substitution, that is, a point mutation. Mutagenic chemicals can damage DNA in other ways such as by methylating a nucleotide. When a cell undergoes division, the methylated nucleotide may be misread and incorrectly copied into the DNA of the daughter cells. Any mutation will be carried in all future progeny of that cell.

Radiation is another source of damage to DNA, usually causing single strand breaks. These breaks are corrected by enzymes in the cell that are part of the DNA repair system. If the DNA repair system is ineffective or if the breaks involve both strands of the DNA molecule, the damage will not be fully corrected. The repair mechanism may itself introduce a mutation in an attempt to repair a DNA break.

The vast majority of point mutations will have no consequence, as they occur either in the introns that do not correspond to any critical cell function or in genes not necessary to the functioning of the cell in which the mutation occurred. Some mutations are lethal and kill the cell, removing it from the body. Only infrequently will a mutation involve a growth control gene, resulting in cancer.

Gene Deletion

In addition to point mutation in the DNA, damage from chemicals or radiation can sometimes cause loss of a large amount of genetic material. The deletion of all or part of a gene may upset the normal regulation of cell growth, resulting in a malignant cell. For example, if a gene that suppresses cell growth is deleted, the consequence may be continuous, unregulated growth. Gene deletion can occur from mistakes in the DNA repair system after extensive damage. If the DNA strands are broken in several places, the repair enzymes, while attempting to join up the cut ends, may erroneously leave out a piece of the genetic material. Additionally there is the phenomenon of genetic instability in tumor cells. For reasons that are not yet known, tumor cells are very prone to rapid accumulation of even more damage to their DNA. As more damage occurs, usually by gene deletions, the tumor evolves and becomes more aggressive. We will see this demonstrated in the case of colon cancer.

Chromosomal Translocation

Another mechanism of carcinogenesis is chromosomal translocation. Broken ends of the DNA from two different chromosomes may be joined incorrectly, resulting in pieces of each chromosome being exchanged. This process disrupts the genes near the breakpoint on each chromosome, with the potential for loss of gene regulation. One of the mechanisms for chromosomal translocation is similar to that for gene deletion. DNA damage resulting in multiple double-strand breaks is incorrectly repaired resulting in splicing of the wrong ends of two different chromosomes.

However, another cause for chromosomal translocation has recently been discovered. In lymphocytes, chromosomal translocation can result from a mistake in the normal process of gene rearrangement. Lymphocytes, as they develop, rearrange their immunoglobulin genes. (See Chapter 7.) During this cutting and splicing of DNA, occasional errors are made that result in chromosomal translocation. In particular, lymphocytes have a predilection for mistaking an oncogene with a normal immunoglobulin gene. For example, during rearrangement of the immunoglobulin heavy-chain gene, the joining segment (Jh) on chromosome 14 may erroneously fuse with the c-*myc* oncogene from chromosome 8. When this t(8;14) chromosomal translocation occurs the abnormal cell grows to a malignant Burkitt's lymphoma (see discussion later in this chapter). To our knowledge, lymphocytes are the only cells in the body that physiologically undergo gene rearrangement. In all other cells, we do not know of any normal condition in which genes move around on the chromosomes. The gene rearrangement process in lymphocytes seems to be a potential source for specific chromosome breaks associated with lymphoma and leukemia.

Infection of a cell by a virus is another mechanism that can result in the activation of an oncogene. Some viruses insert a DNA copy of their genetic material into a human chromosome. This is particularly true for retroviruses. (See Chapter 5.) Viruses may contain a viral oncogene that will stimulate unregulated cell growth. Alternatively, the virus may contain a promoter, which, if inserted near a cellular oncogene, will have the same effect. Viruses have been directly implicated in several animal cancers. The role of viruses in human carcinogenesis is incompletely known.

DNA Repair

There are rare human hereditary diseases in which the DNA repair mechanism is faulty. This accelerates the process of carcinogenesis. Patients with defects in DNA repair, in addition to multiple medical problems, also have a high incidence of tumors. The reason for the frequent appearance of tumors in these patients appears to be that they fail to correct DNA damage and as a result accumulate malignant cells at a much higher rate than normal. Table 8.1 lists some of the known congenital medical syndromes that are associated with defects in DNA repair.

An example from this list is the disease xeroderma pigmentosum. This hereditary disease is due to an inability of the cell to repair damaged DNA. The cell cannot cut out DNA adducts, where damage has occurred. Without excision of these adducts, repair of the strand does not occur. Individuals with xeroderma pigmentosum accumulate DNA damage in their skin from ultraviolet light and have a very high incidence of skin cancers starting in childhood. Ultraviolet light produces a great deal of cell killing in these patients, resulting in severe, sunburn-type lesions. The unrepaired DNA damage that occurs in sun-exposed skin cells usually results in the death of these cells. A few damaged cells survive with mutations at growth control genes. Instead of dying, these cells proliferate as malignant squamous cell carcinomas.

Oncogenes

Oncogenes are growth control genes present in the human genome as well as in the genome of virtually all multicellular organisms. The name onco-gene was given to this group of genes because they were discovered initially in cancer cells. Incorrect expression of an oncogene results in unregulated cell growth, as seen in malignant cells. Oncogenes are, however, a necessary part of normal cell growth. For example, many oncogenes code for cell-surface receptors to growth factors. It is only when the function of an oncogene becomes unregulated that cancer results.

Oncogenes have an important role in normal cell growth functions. Many oncogene products are part of the cell's signal transduction pathway. This pathway modulates the cell's response to external signals. Figure 8.1 delineates the elements (numbered 1 through 4) of the signal transduction pathway. A molecule present in the extracellular environment, such as a growth factor (1), is recognized by a specific receptor (2). When the growth factor binds to its receptor, the molecular conformation changes and a signal

Table 8.1. DNA repair defects — hereditary syndromes.

Disease	Genetics	Defect	Associated cancers
Xeroderma pig-mentosum	Autosomal recessive	Excision repair	Skin
Ataxia telangiectasia		Removal of DNA cross links	Lymphomas
Bloom's syndrome	Autosomal recessive		Leukemias
Fanconi's anemia	Autosomal recessive	Fragile chromo-somes	Leukemias

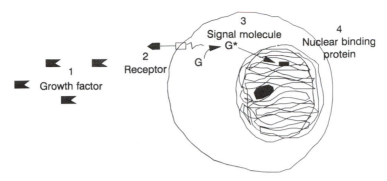

FIGURE 8.1. The signal transduction pathway demonstrates multiple steps in the regulation of gene expression by growth factors.

protein is generated on the inside surface of the cell membrane (3). These signal proteins may be an intracytoplasmic piece of the receptor molecule or another molecule activated just inside the membrane. Tyrosine kinase is an example of a signal molecule. The first signal may serve only as an intermediary, affecting a change in a second messenger. The G proteins are a group of intracytoplasmic second-messenger molecules. G proteins activate yet another molecule that binds to DNA in the cell nucleus (4). This binding to DNA is the end of the signal transduction process and results in the specific gene being turned on or off.

The protein products of oncogenes serve many of the functions along this pathway. Some classes of oncogenes code for proteins that are either growth factors or growth factor receptors. Other oncoproteins are transmembrane signal molecules such as tyrosine kinase. Still other oncogenes produce proteins that modulate the G messenger proteins. And some oncoproteins bind directly to the nucleus, simulating the last step in the signal transduction process. Table 8.2 gives a few examples of oncogenes that fill each of these functions. Of the nearly 100 oncogenes that have been discovered to date, most are involved in this cell transduction pathway.

Signal transduction at each step from growth factor through binding of a regulatory protein to DNA in the cell nucleus is modulated by the products of oncogenes. It is easy to appreciate the importance of the signal transduction pathway for maintaining normal physiologic cell function. Conversely it is easy to see how a mutated oncogene involved in signal transduction could greatly affect a cell's growth potential. The following three examples demonstrate ways in which a mutated oncogene can result in neoplastic growth: (1) An oncogene that codes for a growth factor is mutated so that it fails to respond to negative feedback. The growth factor produced by the oncogene stimulates adjacent cells. The mutated oncogene continues to make growth factor even when normal regulatory feedback would want to

Table 8.2. Oncogenes in human cancer.

Oncogene	Function	Cancer
Growth factors		
int-2	Growth factor	Breast cancer
sis	Platelet-derived growth factor	
Receptors		
Her-2/neu	Protein kinase	Breast cancer
erbB2		
fms	Growth factor receptor	Leukemia
Intracellular signaling		
abl	Tyrosine kinase	Chronic myeloid leukemia
N-ras	Signal transduction	Many cancers
Nuclear proteins		
fos	Nuclear protein	Leukemia
c-myc	Nuclear protein, initiate cell cycle	Burkitt's lymphoma

shut it off, and the growth of adjacent cells continues as a neoplasm. (2) An oncogene that codes for a growth factor *receptor* is mutated and overexpressed. The cell has far too many receptors and is thus ultrasensitive to external stimuli. Under stimulation the cell with too many receptors divides and grows to become a clonal neoplasm. (3) A mutated oncogene makes a protein that binds to DNA and stimulates cell division. The protein that stimulates cell division is expressed continuously, independent of negative feedback, and the cell proliferates as a clonal neoplasm.

Each of these examples results in a neoplastic proliferation of cells producing a tumor. However, each example produces a different grade of tumor along the spectrum from benign to malignant. In the first case, the mutated cell causes other adjacent normal cells to proliferate by production of a growth factor. The proliferating cells do not themselves contain the mutated gene. Their proliferation is polyclonal in that every cell is not an identical copy. This polyclonal proliferation will result in a benign tumor that grows only as large as the supply of growth factor from the mutated cell permits. If the mutated cell were killed, all the benign proliferating cells would stop growing and the tumor would regress. If the tumor is excised, it will not regrow.

In the second case, the mutation results in too many growth factor receptors on the abnormal cell. If stimulated, the mutant cell will divide. The progeny of the mutated cell will also have too many receptors, so they will also divide. This results in a clonal proliferation of identical mutated cells. The tumor grows, but its growth is still dependent on the external stimulus to the growth receptors. If the source of the growth factor to which the cell is ultrasensitive is removed, tumor growth will stop. If the growth factor is a hormone, blockage of hormone secretion will stop the growth of the tumor.

In the third case, the mutation results in unregulated production of a protein that directly stimulates cell division in the mutated cell. The mutated cell proliferates as a malignant clonal neoplasm. The process is autocrine and self-perpetuating. It is not possible to stop tumor growth by blocking extracellular factors. Excision of the tumor will stop further growth only if all mutant cells are removed.

These examples demonstrate how a mutated oncogene can result in a tumor. The clinical course and therapeutic options for each of the three examples are different because of the degree of dependence of the neoplastic growth on outside controls. An understanding of oncogene function in normal cell physiology and in tumors makes it possible to consider new treatment modalities for cancer. If a cell has too many growth factor receptors, a pharmaceutical molecule that blocks these receptors would be useful. If a mutated oncogene is synthesizing a nuclear protein, it would be harder to block its abnormal tumor-promoting function.

A very special property of oncogenes, in addition to their central role in controlling cell growth, is that they exist in two forms. Cellular oncogenes are found, like any other gene, in the DNA of every cell. The second and unique form of an oncogene is called the viral oncogene. In this form, the oncogene is present in the genetic material of a retrovirus. Retroviruses are a special class of viruses that include HIV in humans and numerous cancer-causing viruses in animals. A retrovirus has the property of either replicating in an acute infection or inserting itself into the infected cell's DNA and remaining latent. (See Chapter 5, Fig. 5.2.) The latent retrovirus may be reactivated at a later time.

When oncogenes were first discovered, they were found in the viral oncogene form. The discovery of a gene in a virus that could insert into the DNA of a cell, and later result in that cell becoming malignant, seemed to be the discovery of the cause of cancer. Certainly retroviruses are an important mechanism in the development of malignancy; however, the initial discovery got the story backward. The human genome did not "catch" the oncogene by viral infection; quite the opposite. The retrovirus caught the oncogene from the human cell. Oncogenes in the cellular form are necessary for normal cell growth. Retroviruses borrow a copy of the oncogene and use it to help gain entrance and control of cell division when they infect other cells. A retrovirus actually carries out a very clever genetic engineering step to make sure it takes its own key with it to unlock control of cell division as it moves from one cell to another.

An intriguing feature of oncogenes that was recognized shortly after their discovery is that they are highly conserved throughout evolution. This means that the genomic sequence of the c-*myc* oncogene, for instance, is very nearly the same in a man, a mouse, or a chicken. In general, structural genes such as hemoglobin show substantial variation between animal species. The number of nucleotide sequences for hemoglobin A that are different between man and mouse is about 30%. However, the number of nucleotide sequences that are different between man and mouse for c-*myc* is only 10%. Furthermore, certain portions of the nucleotide sequence for an oncogene may be even more carefully preserved from species to species, with some areas within the gene demonstrating virtually no change. There has been considerable debate about the significance of finding genes with higher-than-expected conservation of sequences between species. One thought is that the gene must be so important that any mutation results in a defect that makes the organism extinct over the time span of evolution. This hypothesis has been proposed for oncogenes that are critical growth control genes in normal physiologic processes. Another hypothesis is that there is some self-correcting aspect to oncogenes, possibly due to an exchange of material with viruses that, over a very long time span, prevents divergence in the nucleotide sequences.

For the *ras* oncogene, the conservation of certain regions of the genome has led investigators to look for some critical function in the portion of the protein encoded by these regions. They found that highly conserved regions encoded certain key amino acids of the *ras* protein that if mutated resulted in a partial loss in enzymatic activity. This loss of activity prevents negative feedback for the down-regulation of *ras*. However, the growth-stimulating activity of the *ras* protein persists, and self-digestion of *ras* protein slows down. This results in a heightened stimulation of cell growth and a tumor.

Tumor Suppressor Genes, Anti-Oncogenes

A newly discovered class of genes called tumor suppressor genes (formerly, anti-oncogenes) has the function of suppressing growth in damaged cells, which inhibits tumors. Although the mechanisms by which these genes halt the development of malignancy are not fully known, some knowledge has been gained. The p53 gene is an excellent example of a tumor suppressor gene. Mutations in the p53 gene are the most common DNA abnormality in tumors, occurring in more than 50% of cancers. This gene codes for a 53-kd protein that binds in the nucleus and at high levels causes the cell to undergo self-destruction, called apoptosis. Normal cells express a very low level of p53 protein. If the cell's DNA is damaged, the expression of p53 increases sharply and the cell commits suicide. However, if the p53 gene itself is damaged or has been mutated earlier on in the history of the cell, this self-destruction cannot occur. The cell will likely still express high levels of the p53 protein, but this will be a mutant protein without the ability to induce apoptosis. Figure 8.2 demonstrates these two pathways of apoptosis (cell death) versus malignancy following DNA damage.

There are many possible mutations that have been found in the p53 gene. Some are inherited, others occur in somatic cells as a mutation in response to some carcinogenic agent. A mutation in the p53 gene (whether a large deletion of base pairs or a point mutation) alters the protein product of the

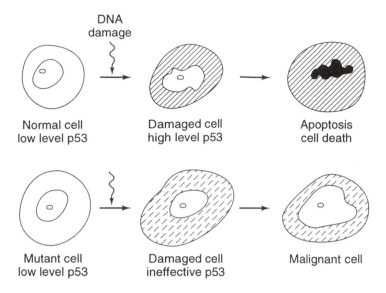

FIGURE 8.2. Schematic model of the alternate pathways following DNA damage in a cell with normal p53 gene and a cell with mutant p53 gene.

gene sufficiently to destroy its capacity to induce a damaged cell to apoptosis. The p53 gene works in an opposing sense to oncogenes such as c-*myc* that stimulate a cell to divide. The protein products of tumor suppressor genes such as p53 exert an antiproliferative effect on cells. Tumor suppressor genes and oncogenes probably function in concert in normal cell growth, alternatively suppressing and encouraging cell division during normal tissue growth and repair. (See Chapter 4 for a discussion of the cell division cycle.)

Another important tumor suppressor gene and one of the first to be discovered is the retinoblastoma gene (Rb), the absence of which is associated with retinoblastoma, lung cancer, and other tumors. People who have inherited a loss of one of the two copies of the retinoblastoma gene have a high incidence of spontaneous development of retinoblastomas. Normally, an individual is born with two copies of Rb. It is now known that some families have an inherited defect in this gene. Heterozygous individuals with only one normal copy of the Rb gene have a higher incidence of retinoblastoma, presumably due to acquired damage to the one remaining copy of the gene. Homozygous individuals with two defective copies of the Rb gene have a virtual certainty of developing a malignant retinoblastoma within the first few months of life. Without any tumor suppressor gene

Table 8.3. Anti-oncogenes or tumor suppressor genes.

Anti-oncogenes	Chromosome	Cancer
Rb	13q	Retinoblastoma, osteosarcoma, small-cell lung cancer
p53	17p	Colon, breast, lung
DCC	18q	Colon
WT1	11p	Wilms' tumor, breast, bladder
NF1	17q	Neurofibroma
?	1q	Medullary thyroid, MEN-2
?	5q	Familial polyposis, colon

function, apparently even very little damage to DNA results in a tumor that cannot be stopped.

Tumor suppressor genes, such as p53 and Rb, do not function as a DNA repair system. They exert their protective influence further along in the process of carcinogenesis, stopping a malignant cell from becoming a tumor. Table 8.3 is a tentative list of some cancers in which loss of function in a tumor suppressor gene may be a contributing factor. This table is very preliminary, as we are just beginning to discover many tumor suppressor genes and their role in cancer. Some of the tumors in Table 8.3, such as retinoblastoma, are rare forms of cancer. Because retinoblastoma is a rare disease, clustering of cases within families was noted. That led investigators to discover the Rb gene. Other tumors in Table 8.3 are not rare, such as lung cancer. Once tumor suppressor genes have been discovered in family clusters of rare cancers, then we have the probes to investigate the role of these genes in more common cancers.

Carcinogenesis is now recognized as a multistep process involving mutation, failure of DNA repair, activation of oncogenes, and loss of tumor suppressor function. The disease we see in patients is the very last stage in this process. With an appreciation of the molecular steps involved in carcinogenesis, we can hope for earlier, more precise cancer diagnosis. Recombinant DNA technology gives us the ability to detect the molecular events associated with cancer. Newer modalities of anticancer therapy are also possible, directed at these molecular defects. In the remainder of this chapter, we will consider several clinical examples, including colon cancer, chronic myeloid leukemia, and Burkitt's lymphoma, that demonstrate the principles of molecular carcinogenesis and their impact on cancer treatment.

Colon Cancer: Model of a Multistep Process

Our understanding of colon cancer in terms of the molecular events of carcinogenesis is more advanced than for most other forms of cancer. Colon cancer develops as a multistep process that progresses with time from benign neoplastic polyps to invasive and finally to metastatic adenocarcinoma. Figure 8.3 demonstrates the stages in this evolution of colon polyp to adenocarcinoma. The molecular alterations that occur as a part of this progression are indicated on the figure. This diagram is redrawn from work by Fearon and Vogelstein (1990), who have contributed greatly to the understanding of the molecular basis of colon cancer.

Let us consider each step in the progression from benign to malignant demonstrated in Figure 8.3. Benign polyps of the colon are increasingly common as the age of an individual increases. Polyps are much more common in Western societies than in other cultures. This increased incidence has been correlated with the amount of fiber in the diet. Sir Denis Burkitt was one of the first to hypothesize that the low-fiber diet of Western societies increases the amount of time that fecal decay products remain in contact with the colon wall. Bacterial breakdown of food products creates natural carcinogens. Other carcinogens may be artificially present in the diet as food additives. With prolonged contact between feces and the glandular epithelium of the colon wall, the potential for carcinogenic events increases. Hyperplastic epithelium and polyps in the colon form with increasing frequency as a function of the patient's age. The DNA of these lesions shows the cumulative effects of lifelong exposure to mutagens. Gene deletions particularly of the long arm of chromosome 5, involving the MCC (missing in colon cancer) gene, are frequently present. Additional changes may include hypomethylation of the DNA.

As the colon lesion progresses from hyperplastic epithelium or early

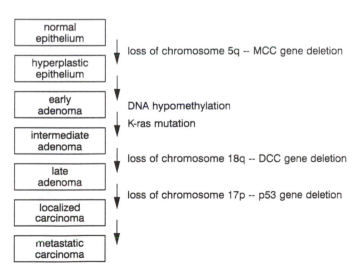

FIGURE 8.3. The molecular steps in the progression of neoplasia of the colon demonstrate accumulation of genetic damage and loss of chromosomes.

stalked benign polyp to a late villous adenoma, more DNA changes are noted, as depicted in Figure 8.3. The DNA from the late stage of polyps, although still classified by pathologists as benign, frequently shows significant DNA damage. Mutation of the *ras* oncogene is common.

The progression from benign to malignant and later to invasive stages of malignant colon cancer is coupled with accumulation of genetic damage. The loss of the DCC (deleted in colon cancer) gene on the long arm of chromosome 18 and the loss of the p53 gene on the short arm of chromosome 17 occurs in 75% of colon cancers. Fearon and Vogelstein (1990) believe that these two genes function either individually or in parallel as tumor suppressor genes. Loss of their function allows progression of a benign tumor to a clinically malignant tumor.

The loss of genetic material in colon tumors is detected by a new technique that employs a large panel of DNA probes (Kern et al., 1989). The battery of probes is designed to detect at least one gene allele on each arm of every chromosome. Southern blotting of tumor DNA hybridized against each of these probes shows loss of portions of chromosomes in all tumor. Adjacent normal colon tissue taken from the edge of the surgical resection of these tumors does not show chromosome loss. This proves that the genetic defects are acquired somatic cell mutations and not inherited. Deletions of chromosomes 17 and 18 are by far the most common site for genetic loss. The p53 and DCC genes are located in the midpoint of these "hot spots" for deletions. Thus p53 and DCC become candidate anti-oncogenes or tumor suppressor genes because their absence is associated with progression to a malignancy. Among a large group of colon tumors, those with p53 or DCC gene deletions have a higher incidence of metastases.

The analysis of genetic loss due to chromosome deletions detected by a large battery of DNA probes has been called by Vogelstein an *allelotype* of the tumor and the extent of deletions is expressed as a fractional allelic loss. The allelotype, although very labor intensive with current manual methods of recombinant DNA technology, has the potential to become a standard tool for analysis and characterization of tumors. The allelotype can be done on solid tumors without the necessity of preparing a suspension of living mitotic cells, which is required for a chromosome karyotype.

The model of colon cancer detailed in Figure 8.3 shows a stepwise progression from benign to malignant with corresponding irreversible damage to the cellular DNA. The DNA damage is only found in the diseased tissue; DNA in normal colon tissue adjacent to these lesions is undamaged. This model greatly increases our understanding of human cancers. The development of a cancer is a result of the balance between carcinogenic events and failure to repair DNA damage. It is a multistep process that is probably occurring in all individuals. Those who develop cancer have undergone the evolution in more rapid fashion. This model explains why colon cancer has a higher incidence with advancing age.

The diagnosis of colon cancer and the assessment of risk from benign polyps will be greatly changed by the molecular view presented in this model. Pathologists, in addition to assessing the grade or stage of a tumor based on morphologic evidence, will look at the molecular lesions. Molecular diagnostics may add more precision to prognostic judgments. It may be possible to screen for cells shed in the feces to look for evidence of DNA damage. Classical cytology is not possible on cells shed from the lower gastrointestinal tract as these cells rapidly break down along with decaying food products. However, DNA is more stable, and with methods such as PCR, it is theoretically possible that small numbers of cells—for example, with a deletion in the p53 gene—could be detected in a screening process using stool samples.

Growth factors given as a drug might provide the regulatory signals missing as a result of defective genes. Growth factor therapy could halt or even reverse the progression of colon cancer. For example, the p53 protein could be given as a therapeutic agent in tumors associated with loss of normal p53 gene function. A molecular understanding of the various stages in cancer development and progression is the basis for molecular therapeutics.

Lymphoma, Leukemia

The molecular biology of lymphoma and leukemia is, like colon cancer, well advanced, and has already led to new therapies. The easy accessibility of blood cells for analysis is a major reason why molecular studies got a headstart in hematology. The special property of immune gene rearrangement in lymphocytes has given us a marker for the molecular characterization of malignant lymphocytes. Identification of a unique rearrangement allows us to demonstrate that all lymphomas are a clonal expansion of a transformed lymphocyte arrested in its maturation. For Burkitt's lymphoma we have a fairly good idea as to how this neoplastic transformation comes about. Another hematologic malignancy, chronic myeloid leukemia, was the first tumor for which a chromosomal translocation was discovered as a marker of the malignant cells. It was also the first disease in which a breakpoint cluster region, the molecular counterpart of a chromosomal translocation, was discovered. Now all leukemias are known to be associated with specific chromosomal translocations. The molecular events on the broken chromosomes are rapidly being discovered and cloned. Use of growth factors in therapy of lymphoma and leukemia has evolved as a result of these molecular discoveries. Lymphoma and leukemia are outstanding examples of the impact of molecular biology on medicine.

The maturation of lymphocytes from stem cell to mature B or T lymphocytes is accompanied by a physiologic rearrangement of the immune genes. This process was detailed in Chapter 7. A problem with the rearrangement of the immune genes is that this complex process is an invitation for error.

Even when things go right, only one lymphoid stem cell out of three success-fully rearranges its genes and becomes a mature lymphocyte. One stem cell successfully undergoes rearrangement and subsequent clonal expansion, whereas two others fail to carry out the process and presumably are de-stroyed through programmed cell death. The development of lymphoma as a malignant proliferation of unregulated lymphocytes occurs as an error in the gene rearrangement process. Most of the various types of lymphoma have now been associated with a molecular error in gene rearrangement. The error is invariably a fusion between the normal immune gene sequences and an oncogene. For various lymphomas, different oncogenes are in-volved, but the end result is always fusion of an immune gene plus an oncogene.

Burkitt's Lymphoma

An aggressive form of lymphoma in African children was first described by Burkitt and O'Conor (1961). Sir Denis Burkitt used epidemiologic observa-tions to discover the association between cancer and environmental factors long before a theoretical or molecular basis for cause and effect was known. He also described the role of high-fiber diets in lowering the incidence of colon cancer. Burkitt's lymphoma has a number of features that raise very important questions in view of today's molecular approach to disease. Bur-kitt's lymphoma is a disease endemic to certain geographic areas of Africa that are hot, humid lowlands with a very high level of malaria. Burkitt's lymphoma in Africa most frequently presents as a tumor of the jaw. Virtu-ally 100% of patients with African Burkitt's lymphoma have evidence of persistent Epstein-Barr virus (EBV) infection as demonstrated by the pres-ence of the EBV genome in the tumor cells and increased antibody titers to viral proteins. The exact role of EBV in Burkitt's lymphoma is only now being discovered, despite the fact that this was the first human cancer in which a virus was implicated. There are some puzzling features that made understanding the relationship between the disease and the virus difficult. Epstein-Barr virus is widely distributed over Africa, but Burkitt's lym-phoma is restricted to areas with endemic malaria. Burkitt's lymphoma is rare outside of Africa but does occur as the so-called sporadic or non-African Burkitt's lymphoma. In the United States, Burkitt's lymphoma is a rare tumor of children and usually presents as abdominal lymphadenopathy or leukemia and rarely involves the jaw. In non-African cases, only 15–20% of patients have evidence of EBV infection, which is not too different from the background level of infection in the general population. Neverthe-less, both African and non-African Burkitt's lymphoma when studied by molecular techniques demonstrate a t(8;14) translocation with the insertion of the c-*myc* oncogene into the immunoglobulin gene.

Quite recently an additional form of Burkitt's has arisen that gives fur-ther insight into the pathogenesis of this disease. Patients with AIDS de-velop a number of tumors, of which Burkitt's lymphoma is one of the more common. In these patients, EBV genome is frequently found in the tumor. AIDS patients with Burkitt's lymphoma also have translocation of c-*myc* into the immunoglobulin gene.

Taken together, the accumulated observations on Burkitt's lymphoma have led to a theory for its pathogenesis. Figure 8.4 schematically demonstrates this model. Epstein-Barr virus is attracted to receptors on B lymphocytes. When the virus attaches to a surface receptor, the lymphocyte is stimulated to divide. A polyclonal proliferation of B lymphocytes follows acute viral infection. Usually the proliferation of B lymphocytes is controlled by the T lymphocyte part of the immune system. The T lymphocytes will, after a period of some months, suppress the proliferating infected B lymphocytes. In patients who are immunocompromised by malaria or AIDS, T lymphocyte activity is impaired. The proliferation of B lymphocytes as a polyclonal expansion continues without control by the impaired T cells. These proliferating B lymphocytes are subject to a high rate of error in gene rearrangement.

Either internal errors or external carcinogens lead to the next step. The c-*myc* oncogene from chromosome 8 translocates into the heavy-chain-joining region of the immunoglobulin gene (Jh) on chromosome 14. This results in Jh erroneously fusing with the c-*myc* oncogene. The fused c-*myc*/Jh gene falls under the control of the immunoglobulin gene promoter. As the lymphocyte matures, immunoglobulin synthesis is turned on by the

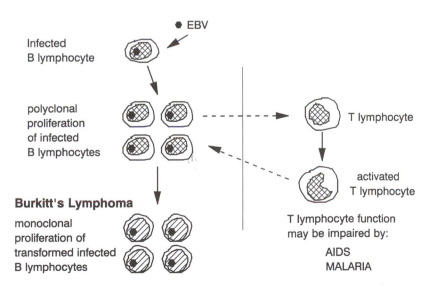

FIGURE 8.4. A model for the relationship between Burkitt's lymphoma and infection with Epstein-Barr virus shows the multiple steps including infection, impairment of the T lymphocytes, and final malignant transformation leading to a tumor.

promoter for this gene. In the malignant lymphocytes with c-*myc*/Jh fusion, the c-*myc* oncogene protein is produced along with an abortive attempt at immunoglobulin production. The c-myc protein binds to DNA within the nucleus and stimulates further cell division. The result is arrested maturation at an early lymphoblastic stage with self-stimulation for further proliferation.

This type of aberrant oncogene function is the same as example #3 in the description of oncogenes at the start of this chapter. As we predicted then, a malignant cell that is totally self-stimulatory results in an aggressive malignancy. Burkitt's lymphoma with continuous expression of a deregulated c-*myc* gene is one of the fastest-growing tumors. This model also explains the high incidence of Burkitt's lymphoma in African and AIDS-associated cases where impairment of the T lymphocyte arm of the immune system is a common event.

Follicular Lymphoma

In follicular lymphomas, the story is virtually the same as in Burkitt's lymphoma. A lymphoid stem cell, in an attempt to rearrange its immunoglobulin genome during maturation, erroneously fuses with an oncogene. This time the oncogene, *bcl-2*, is from chromosome 18. *bcl-2*, like c-*myc* in Burkitt's lymphoma, fuses with the Jh portion of the immunoglobulin gene. The bcl-2 protein becomes expressed under the promoter for the immunoglobulin gene. The result is a clonal expansion of a malignant lymphocyte, but one that is more mature than the characteristic lymphoblast in Burkitt's lymphoma. Follicular lymphoma with *bcl-2*/Jh fusion is an indolent low-grade malignancy.

The function of the bcl-2 protein has recently been discovered and sheds some light as to how bcl-2 expression promotes a malignant tumor. The bcl-2 protein binds to the inner surface of the mitochondrial membrane and prevents programmed cell death (Hockenbery et al., 1990). Normally, a lymphocyte that does not successfully carry out gene rearrangement dies. Abnormal bcl-2 expression appears to prevent programmed cell death in a lymphocyte. A transgenic strain of mice bearing an inserted *bcl-2*/Jh fusion gene has been created as a model for lymphoma. These mice develop lymphoid hyperplasia that progresses rapidly to lymphoma (McDonnell and Korsmeyer, 1991). Both follicular lymphomas and high-grade Burkitt-type lymphomas have been seen in these transgenic mice. These results are tantalizing evidence that deregulation of bcl-2 expression prevents the filtering out of lymphocytes with abnormally rearranged genes.

Despite our recently acquired understanding of how lymphomas occur as an error in gene rearrangement, we must admit that there is still much to be learned. The *bcl-2*/Jh fusion is occasionally seen in aggressive lymphomas as well as in the low-grade follicular form (Gulley et al., 1992). The molecular event is the same in high-grade or low-grade lymphomas, yet these lymphomas behave quite differently. Continuing intense research into the molecular biology of lymphoma is likely to provide more answers and hopefully lead to new therapies in the near future.

Molecular Diagnosis and Staging of Lymphoma

Tumors are, by definition, a clonal proliferation of a single abnormal cell. Therefore, one of the most important criteria in the difficult diagnostic decision between benign and malignant lymphoid neoplasia is evidence of a monoclonal proliferation of lymphocytes. By microscopy, a clonal proliferation is recognizable as a homogeneous population. Using immunocytochemistry, monoclonality is recognized when lymphocytes with a single surface marker, such as IgM kappa, are a predominant subpopulation. However, malignant lymphoma is a complex process involving not only the proliferation of a monoclonal tumor population but also an intermixing of reactive polyclonal lymphocytes, both T and B cells within the tumor. This intermixing of polyclonal reactive cells can mask the morphologic features of lymphoma. Immunocytochemistry can identify the tumor cells only when they are a clearly dominant fraction. The difficulty in the diagnosis and classification of lymphomas by microscopy alone is demonstrated by the long and tangled history in the naming and classification of lymphomas.

In the diagnosis of lymphoma DNA probes can assist because of their ability to demonstrate unequivocally even a minor monoclonal population of lymphocytes. The abnormal clone produces a different banding pattern that germline cells on a Southern blot (see Fig. 7.5) or by PCR. This abnormality in the DNA of the malignant clone can be a rearrangement of the immunoglobulin or T-cell-receptor genes, a chromosomal translocation, or the presence of viral DNA integrated into the lymphocyte genome. For establishing the clonality of lymphoid neoplasia, DNA probes are the most sensitive method, identifying as few as 5% monoclonal cells in a Southern blot. Furthermore, the type of genetic abnormality helps classify the tumor. Immunoglobulin gene rearrangement is characteristic of B cell lymphomas, whereas T-cell-receptor-gene rearrangement marks T cell lymphomas.

A clinical example of the molecular diagnosis and staging of lymphoma will demonstrate the usefulness of molecular diagnostics. Follicular lymphomas are characterized by a translocation between chromosomes 14 and 18 in which the bcl-2 oncogene on chromosome 18 breaks at a very specific point and inserts within the immunoglobulin gene on chromosome 14. The fusion of the bcl-2 oncogene with the immunoglobulin gene is thought to inhibit the death of B lymphocytes resulting in the accumulation of small, cleaved cells characteristic of follicular lymphoma. To detect whether a translocation has occurred, DNA is extracted from a lymph node, blood, or bone marrow and analyzed by PCR. Two primers, one that binds to a site on chromosome 14 and the other on chromosome 18, are mixed with the DNA, and a PCR reaction is carried out. If there are pieces of DNA in the sample in which chromosomes 14 and 18 have been broken and rejoined, then the two primers will be close together and the PCR reaction will produce an amplified product. If there is no translocation between chromosomes 14 and 18, the two primers will be very far apart and amplification does not occur. (See Chapter 3 for an explanation of PCR.) Without amplification, no specific band is seen on the agarose gel electrophoresis. Figure 8.5 shows a gel electrophoresis of the products of a PCR reaction using primers for Jh on chromosome 14 and bcl-2 on 18. Each lane represents a different patient sample. Two lanes (4 and 9) show an extra band marked by arrows. These two patients have the t(14:18) translocation. The single band that is common to all lanes is the product of the amplification of another gene that is done as a positive control to demonstrate that the PCR reaction worked in all patient samples. The PCR method can be carried out more rapidly than a Southern blot and uses very small amounts of DNA. PCR does not require use of a radioactive probe.

An additional technical point that is worth emphasizing is the special sample characteristics that help make these methods attractive to the clinical laboratory. The Southern blot requires only 10 μg of DNA (easily isolated from 1 million cells). PCR requires less than 1 μg and in fact can be made to work on even a single cell. For PCR analysis, simple lysis of the cells is sufficient preparation of a clinical sample. Extraction and purifica-

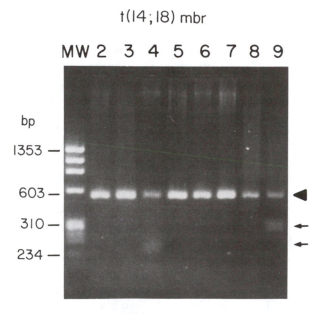

FIGURE 8.5. An agarose gel electrophoresis of the products of a PCR reaction to detect the t(14;18) chromosomal translocation shows a positive result in patient samples 4 and 9 (arrows). The large arrowhead marks the product of a normal gene, amplified as a control.

tion of the DNA from the sample, which is required for Southern blot analysis, is not necessary for PCR. DNA is a very stable molecule and can be extracted from samples maintained at room temperature for several days. DNA suitable for PCR has even been successfully isolated from paraffin-embedded surgical pathology biopsies that are years old. One serious drawback to PCR technology is the danger of contamination from the amplified DNA product back into new samples. Since the amount of amplified DNA is 1-millionfold greater than that present in the starting sample, the most insignificant carryover of amplified DNA products back to the preparation area will result in false positives on future tests.

Detection of Minimal Residual Disease

The great sensitivity of the PCR technique, with the ability to detect one in 10^5 abnormal cells, has opened the door to a new means of staging lymphoma and detecting submicroscopic disease. A biopsy of the bone marrow typically must have 5% or greater lymphoma cells for the process to be recognized by morphologic examination. This 5% limit also represents the sensitivity of Southern blot analysis of DNA. The PCR extends the sensitivity to 0.01% lymphoma cells infiltrating marrow, blood, or other tissues. Molecular restaging of blood, marrow, and other sites in patients with lymphoma will undoubtedly result in many patients being classified to higher-stage disease. The clinical significance of submicroscopic involvement of marrow and blood has yet to be determined, and, in fact, we may learn that submicroscopic involvement of all lymphoid and hematopoietic organs is common in lymphoma. PCR analysis of patients in remission after systemic chemotherapy offers an ultrasensitive laboratory method to detect minimal residue disease.

T Cell Malignancies

Molecular studies are even more important for the diagnosis of T cell lymphomas than for B cell neoplasms for several reasons. We do not have as good a panel of immunologic markers for T lymphocytes as we do for B. Immunocytochemistry is rarely sufficient to prove monoclonality of a T cell proliferation. Additionally, T cell malignancies frequently occur as infiltrates in extranodal sites where the malignant cells make up only a small population of the biopsy. Skin infiltrates are a particularly tough diagnostic problem. Southern blot analysis for rearrangement of the T-cell-receptor genes offers an important tool with which to overcome these difficulties. Figure 8.6 demonstrates a Southern blot showing gene rearrangement of the T-cell-beta-chain receptor. Lanes 1, 2, and 3 are DNA isolated from a suspected cutaneous lymphoma, each digested with a different restriction enzyme. The germline pattern of unrearranged DNA is shown in the adjacent negative control lanes marked with a minus sign. We can see the normal band plus an additional band resulting from rearrangement of the T-cell-receptor gene in the digested DNA present in lanes 1 and 2, but not

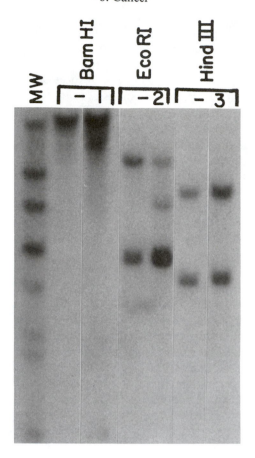

FIGURE 8.6. A Southern blot analysis of DNA from a cutaneous infiltrate of lymphocytes shows a clonal rearrangement of the T-cell-receptor gene.

3. Depending on how the DNA is rearranged, one of the three different digests may not show the rearrangement. This Southern blot analysis indicates that a clonal population of T cells consistent with a malignant infiltrate is present in the tumor biopsy.

DNA probe technology, in addition to recognizing rearrangement of immunoglobulin and T-cell-receptor genes, is also an excellent tool for detecting chromosomal translocations. Karyotypes on lymphomas, although diagnostically valuable, have been difficult to perform and are not routinely carried out in most hospitals. The use of DNA probes may now give this capability to a wide range of clinical laboratories. The application of DNA probe technology to the diagnosis and monitoring of lymphomas comes at a very useful time with respect to new therapies being applied to the treatment of lymphoma. Biological response modifiers such as interferons, interleukins, tumor necrosis factor, and monoclonal antibody therapy have dramatic impact on the morphology and immunophenotype of malignant lymphomas. The tumor cells may mature when exposed to biological response modifiers, causing a marked change in microscopic appearance. This change in morphology and immunologic phenotype of the malignant cells can mask the presence of the lymphoma if the biopsy is examined with only conventional techniques. The genetic makeup of the tumor cells is, however, not altered by biological response modifier therapy. DNA probes for immunoglobulin gene rearrangement, or probes for chromosomal translocations, will reveal an abnormal monoclonal population irrespective of phenotypic changes resulting from growth factor therapy. Thus, DNA probe technology becomes a valuable tool in monitoring patients on special therapies.

Chronic Myeloid Leukemia

The molecular biology of chronic myeloid leukemia (CML) is another fascinating example of the rapid increase in both knowledge and clinical usefulness resulting from the application of recombinant DNA technology to the study of malignancy. The whole story from the initial basic-science discovery of the underlying genetic event in CML to widespread clinical application of this knowledge spans less than 5 years. In the early 1980s, a group of researchers in Rotterdam asked the question, "When chromosomes break in tumors, do the breaks always happen at the same place on the DNA molecule, or are the breaks widely distributed?" They chose to study CML to answer this question because CML is characterized by a specific cytogenetic abnormality named the Philadelphia chromosome. The Philadelphia chromosome, the first cytogenetic abnormality recognized as a tumor marker, is formed by a reciprocal translocation between the long arms of chromosomes 9 and 22. In the vast majority of patients with CML, the Philadelphia chromosome can be seen in mitotic cells prepared from the blood. The Philadelphia chromosome is an acquired abnormality present only in the leukemic cells. Other cells in these patients do not show the abnormality.

Knowing this, the Rotterdam group thought CML would be a good model in which to study their question as to whether the breaks associated with a chromosomal translocation were at a single fixed point or spread out over a large distance. They took 11 patients with CML and began the laborious task of mapping where the breakpoint occurred on chromosome 22 for each patient. This is now an easy task thanks to their discovery and the development of probes for chromosome 22, but in 1983 this was a difficult job.

The result of the Rotterdam group's investigation was exciting (Heisterkamp et al., 1985). They discovered that the breaks on chromosome 22 in all 11 patients were clustered in a very small region which they named the breakpoint cluster region (*bcr*). This was the first example of a chromosomal translocation shown to have a small *bcr*. Another group of researchers in England asked nearly an identical question. They too chose a group of CML patients and began mapping the break in the chromosomes, only they chose chromosome 9 rather than 22 as the place to begin. They did not find a clustering of breakpoints on chromosome 9. The fact that the breaks are clustered on chromosome 22 but not on 9 suggests that if there is a molecular basis for CML, the action is likely to be on chromosome 22. The Rotterdam group also noted that the part of chromosome 9 that was fused with chromosome 22 was the Abelson (*abl*) oncogene. The involvement of an oncogene seemed to them an exciting hint that this event might be the cause for CML.

After the discovery of the *bcr* region on chromosome 22, further understanding of the molecular biology of CML was rapid. The *bcr* region was discovered to be within a gene that has subsequently been named BCR in recognition of the fact that this gene was the first in which the phenomenon of breakpoint clustering had been discovered. The function of this BCR gene on chromosome 22 is as yet unknown. We are sure that BCR is a gene, however, because of its intron/exon structure, its open reading frame, and its ability to serve as template for mRNA. When the BCR gene is interrupted by the insertion of *abl* from chromosome 9, a new fusion gene called BCR/*abl* is generated. In this fusion gene, the sequences of *abl* are preserved and a small portion of BCR sequences is added. The fusion of the two genes following the chromosomal translocation also results in the gene being turned on. Figure 8.7 demonstrates these molecular events underlying the Philadelphia chromosome translocation characteristic of CML.

The clones of leukemic blood cells that develop from the single transformed stem cell in which the chromosomal translocation occurred all carry the BCR/*abl* gene and express the mRNA from this gene. The *abl* gene normally codes for a tyrosine kinase; its mRNA is 7.0 kb long and the protein it generates has a molecular weight of 145 kd. When BCR and *abl* are fused, the gene is longer and the mRNA is 8.5 kg with a protein of 210 kd. This BCR/*abl* fusion tyrosine kinase operates like the *abl* protein but accepts a wider range of substrates and is kinetically faster. All the blood cells that are part of the leukemic clone express this fusion protein. CML is a neoplasm similar to our example #2 at the start of this chapter. The mutated tyrosine kinase is an ultrasensitive receptor molecule.

To demonstrate whether this fusion gene is truly the cause of leukemia, Daley et al. (1990) removed the fusion gene from a sample of leukemic cells and inserted this gene into the embryo of a mouse. The transgenic mice produced in this fashion have cells that contain the abnormal BCR/*abl* fusion gene. These mice invariably develop leukemia within the first months of life. Thus, the BCR/*abl* event resulting from the Philadelphia chromosome translocation appears to satisfy Koch's postulates for the cause of the

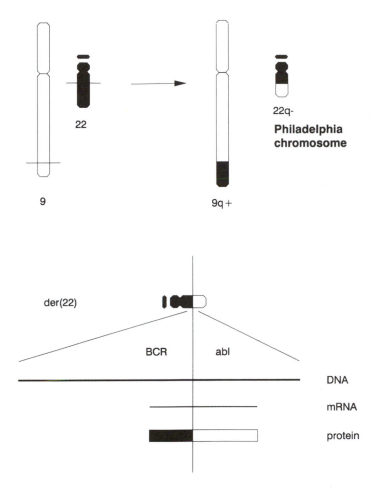

FIGURE 8.7. In CML, the t(9:22) translocation creates an abnormal Philadelphia chromosome (a) that at the molecular level (b) is a fusion of the BCR and *abl* genes. The BCR/*abl* fusion results in a hybrid mRNA and protein containing elements of both genes.

disease: The presumed etiologic agent, the BCR/*abl* fusion gene, can be isolated, and when placed into the appropriate environment, it reconstitutes the disease.

The treatment of CML patients with chemotherapy, such as with alpha 2 interferon, in some patients results in the rapid suppression of the BCR/*abl* gene expression. The abnormal 8.5-kb mRNA disappears from the cells. Sequential Northern blots made from leukemic cells after interferon therapy show the disappearance of the gene product within hours after drug therapy is started. Continuing drug therapy with alpha 2 interferon results in the gradual shift from a leukemic to a normal blood film over a period of months. The typical high white blood count falls to a normal range of under 10,000/μl. About 60% of patients treated with interferon enter a hematologic remission. Sequential DNA studies on the blood cells from these patients show that many demonstrate a progressive, slow disappearance of the leukemic cells as marked by the BCR/*abl* gene. In other CML patients, however, the interferon therapy seems to suppress the leukemic behavior of cells, presumably by shutting off the BCR/*abl* gene, but the leukemic cells remain in the bone marrow. Thus, many patients in hematologic remission still have 100% leukemic cells as measured by Southern blots. Bone marrow transplantation from a compatible donor, when available, is the current, most-successful therapy for CML.

A presumptive diagnosis of CML is still made on the basis of the characteristic findings of a high white blood cell count with left shift and a large spleen. But the diagnosis is now confirmed by molecular Southern blot analysis of DNA from the blood cells. Southern blot analysis is less expensive and more rapid than the traditional karyotypic analysis (Ayscue et al., 1990). Furthermore, Southern blot analysis is more sensitive, picking up about 10% of patients who appear to have a normal karyotype but in fact have the abnormal BCR/*abl* gene fusion. Figure 8.8 shows a Southern blot analysis of a patient with suspected CML. The patient shows germline plus rearranged bands in lanes 1 and 2 with BamHI and EcoRI digests. In lane 4, a HindIII digest, only a germline band is present. Positive and negative control samples are run in adjacent lanes, along with molecular weight markers as part of the quality assurance process. In addition to detecting the Philadelphia chromosome on a Southern Blot, we can also quantitate the percentage of cells that are positive for the translocation. Each leukemic cell has one Philadelphia chromosome and one normal chromosome. A sample of 100% leukemic cells would give two bands of equal density on a Southern blot analysis as demonstrated in Figure 8.8. The normal chromosome produces the germline band; the Philadelphia chromosome is responsible for the rearranged band. By scanning the density of the germline band, relative to the density of the rearranged band, we can calculate a ratio of normal cells to Philadelphia-chromosome-positive cells. If the bonemarrow sample contained only 50% leukemic cells, then the germline band would have a density of 75% and the rearranged band 25%. This is because each normal cell contributes two normal germline chromosomes, while a leukemic cell contributes one germline and one rearranged chromosome. The quantitative Southern blot is necessary to monitor patients on interferon therapy.

Our understanding of CML, although detailed, is not complete. When patients advance from the chronic phase to blast crisis, no consistent change in the molecular rearrangement of the leukemic cell's DNA has yet been discovered. The disease of CML has seen many firsts: the first tumor discovered to have an abnormal marker chromosome and the first abnormal

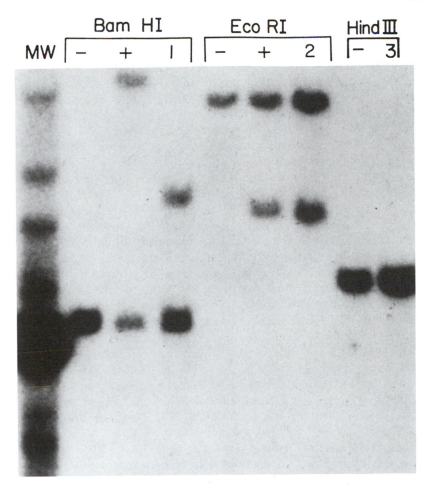

FIGURE 8.8. Southern blot analysis of DNA from a patient with CML demonstrates a BCR/*abl* rearrangement.

chromosome discovered to have a breakpoint. The disease remains a testing ground for further observations and will be, one hopes, the object of new therapies designed to take advantage of our molecular insight into the pathogenesis.

Acute Promyelocytic Leukemia

The story of molecular biology's role in another blood cancer—acute promyelocytic leukemia (APL)—is brief but exciting. Recent discoveries show that this leukemia is characterized by a chromosomal translocation that interrupts the gene for a vitamin A receptor (de The et al., 1990; Longo et al., 1990). What's exciting about that is the immediate implications for therapy. If the unregulated cell proliferation in APL is due to an acquired defect in the cell-surface receptor for vitamin A, then treatment with high doses of this vitamin should override the defect. In fact vitamin A therapy does produce rapid remissions in many cases of APL. The credit for the discovery of this nontoxic antileukemia therapy, however, goes to clinicians and not to molecular biologists.

As is so often the case in the history of medicine, astute physicians had long recognized the special characteristics of APL. The clinical features that distinguish APL from other forms of leukemia are a high incidence of disseminated intravascular coagulation, extreme sensitivity to daunorubicin, and ability to reinduce patients through multiple relapses with the same drug. Observations of APL patients under therapy have repeatedly shown that this leukemia is prone to induction of maturation. Patients may show a rapid change from a marrow packed with promyelocytes to a marrow packed with mature neutrophils as a response to treatment, without an interceding hypoplastic phase. Laboratory studies in cell cultures of APL cells plus the accumulated clinical observations encouraged physicians to treat APL with drugs that would induce maturation. Early attempts included use of butyric acid, dimethylsulfoxide (DMSO), low-dose cytosine arabinoside (ARA-c), and vitamin A. Partial successes were achieved, particularly with low-dose ARA-c. A therapeutic trial by the Chinese of vitamin A, in the all-*trans*-retinoic acid form, showed the most dramatic results (Huang et al., 1988).

The recent molecular studies that demonstrate that APL is characterized by a rearrangement of the vitamin A receptor now provide the data that make the treatment of APL with vitamin A understandable. The molecular research followed what has now become a standard line of reasoning based on the belief that all forms of leukemia are due to an aberrant chromosome rearrangement. Acute promyelocytic leukemia has been known for a number of years to be characterized by a t(15;17) chromosomal translocation in the malignant cells. The presence of a specific chromosomal translocation points to the interruption of a specific gene. The recent studies demonstrating that the rearranged gene in APL is the receptor for *trans*-retinoic acid explain the success of vitamin A therapy. The interruption of the receptor prevents normal feedback inhibition of cell proliferation via this pathway. Studies in vitro show that if APL cells are supplied with *trans*-retinoic acid at a higher concentration than would normally be required to bind to the receptor, the block in the pathway can be overcome. The required doses are within a therapeutic range, and side effects are minimal, certainly when compared with the toxic side effects of standard chemotherapy. Hematologists are excited by an understanding of the molecular basis for this disease and the early therapeutic successes. The demonstration of a molecular biology breakthrough immediately coupled with an advance in therapy justifies some of the enthusiasm in recombinant DNA technology applied to medicine.

Other Solid Tumors

A working hypothesis among cancer researchers is that each form of cancer has specific molecular events that help distinguish it and allow us to understand the biology of the tumor. Hopefully, these molecular events will also be a guide to diagnosis and treatment. The limited data now available suggest that for each tumor we will discover oncogenes that, when activated, stimulate the malignant proliferation of the tumor cells, as well as anti-oncogenes whose loss allows the tumor to progress. A few additional examples of cancers whose molecular events we are only beginning to understand will serve to demonstrate the rapid growth of this area.

Breast Cancer

The molecular biology of breast cancer is incompletely understood. About 10% of breast cancers occur in women who have an inherited loss of function in the *brca1* tumor suppressor gene. The *brca1* gene probably functions to suppress proliferation of breast epithelial cells with damaged DNA. The mechanism is as yet unknown but it is likely to be similar to the p53 tumor suppressor gene discussed earlier. In those patients with a hereditary increased risk of breast cancers, the abnormal *brca1* gene can be detected by screening. A simple blood sample, using the DNA harvested from peripheral blood lymphocytes, is sufficient. In women found to be at increased risk for breast cancer because of a positive *brca1* screen (or a family history), much closer clinical observation would be indicated. The screening of women for inherited breast cancer risk is likely to become one of the first major results of the human genome project (National Advisory Council for Human Genome Research, 1994). Many new concepts will be tested in such an undertaking. Questions to be considered include the following: Exactly what clinical steps will be taken in women positive for *brca1*? How will their increased risk be treated by employers and health insurers? How will a positive screen affect a woman's peace-of-mind?

For the remaining 90% of breast cancers that do not occur in women with an abnormal *brca1*, other molecular events are useful in monitoring the disease. An acquired mutation in the HER-2/*neu* (alternatively called *erb*B-2) oncogene results in an abnormally high expression of HER-2/*neu* protein within the tumor. Patients with increased HER-2/*neu* as determined by immunohistochemical staining of the tumor sample seem to have a poorer prognosis. Dennis Slamon et al. (1989) have presented the most convincing evidence that gene amplification of HER-2/*neu* with resulting overexpression of the protein is a sign of a biologically aggressive tumor. Other molecular parameters, especially DNA ploidy and cell cycle studies on breast tumors, also seem to be significant independent predictors of clinical behavior. Flow cytometry studies on breast tumors (see also Chapter 4) have become standard in the workup of early stage breast cancer.

One of the most difficult problems in the treatment of breast cancer is what to do for early-stage disease. Does the patient with a malignant tumor smaller than 2 cm need more than a lumpectomy? Mastectomy with examination of axillary lymph nodes is now done more for adequate staging than for therapy. Molecular studies such as flow cytometry, *brca1*, or HER-2/ *neu* may provide equal or better staging information. With this information surgical therapy, radiation therapy, and chemotherapy can be optimized for each patient. Large-scale clinical trials are under way to correlate the molecular biology of breast cancer with clinical outcome.

Lung Cancer

The story of the molecular biology of lung cancer is at a stage similar to that of breast cancer. Certain specific mutations have been seen more commonly in lung cancer and appear to be somewhat specific for this disease. Amplification of the L-*myc* oncogene and point mutation of the *ras* oncogene are somewhat specific for lung tumors. Since early-stage lung cancer is rare and most tumors when recognized clinically have a poor prognosis, we may not be seeing the molecular biology of lung cancer in an early-enough stage in its natural history to appreciate the spectrum of changes.

A potentially innovative approach to lung cancer screening may become available by using recombinant DNA technology. Since a specific point mutation in the *ras* oncogene is characteristic of lung cancer, screening for this mutation in sputum might "theoretically" provide for early detection of disease. Just a few bronchial epithelial cells present in a sputum sample constitute enough DNA for analysis by using the PCR technique. If only a small number of cells with *ras* oncogene mutations are present in the sample, they will be detected. It will require a large clinical trial to know whether this approach could really be significant in detecting early-stage disease. Then the question of whether early-stage disease is more treatable would still need to be answered.

Thyroid Cancer

Another example of the unfolding story of the molecular biology of tumors is provided by very recent discoveries in thyroid neoplasia. A new oncogene, *gsp*, has been discovered that seems to be relatively specific for benign and malignant thyroid tumors. The *gsp* oncogene is associated with a "G protein" whose function is to transport signals from hormone receptors on the cell surface to the nucleus. (See Fig. 8.1.) *Gsp*, like so many other oncogenes, is involved in cell growth regulation by providing a transporter signal molecule. One study noted that 100% of multinodular goiters had *gsp* mutations not present in the DNA of normal cells from these patients

(Lyons et al., 1990). In thyroid tumors of higher grade, including adenocarcinomas, the *gsp* mutations were present along with additional oncogene mutations such as in *ras*. Similar to colon cancer, thyroid neoplasms demonstrate the same accumulation of mutations in oncogenes, leading to a molecular evolution from benign to highly malignant tumors.

Cancer of the Uterine Cervix and HPV

Cancer of the uterine cervix is common, but fortunately it is curable if detected early. Molecular discoveries have found that one of the etiologic agents associated with cervical caner is the human papilloma virus (HPV). HPV has not yet been shown to be the cause of squamous-cell cancer of the cervix; more likely it is one of several possible factors associated with a multistep progression from benign to malignant. However, the detection of HPV by molecular hybridization offers an additional screening tool for cervical cancer.

The standard screening method for cervical cancer is cytologic examination of exfoliated cells on a Pap smear. A Pap smear can detect cervical abnormalities at an early, premalignant stage at which limited surgery is curative. Cancer of the cervix in the vast majority of cases develops in a progressive fashion beginning with dysplasia. The progression of cervical dysplasis has been broken down into various classes diagnosed by Pap smears and cervical biopsies to aid in clinical management decisions. Pap smears are graded as normal, atypical, and low- or high-grade squamous intraepithelial neoplasia. Beyond intraepithelial neoplasia is invasive cervical carcinoma, which is not curable by limited surgical resection and requires a combination of radical surgery and chemotherapy for salvage of patients.

In the United States, hundreds of thousands of Pap smears are performed each year. Abnormal results require close clinical follow-up. The intensity of clinical follow-up for patients with abnormal Pap smears has been a subject of significant debate. The majority of cases with mild cervical dysplasia do not progress to invasive cancer. However, a few patients do progress through the various stages of dysplasia more rapidly.

The association of infection with HPV and the development of cervical cancer are now partially understood. The knowledge that HPV has a role in the development of disease is significant in the grading and clinical management of patients. From the presence of koilocytes on a Pap smear HPV infection can be inferred. However, molecular detection of HPV infection is more specific. Two slightly different molecular methods are employed: in situ hybridization or dot-blot hybridization. In situ hybridization applies to nucleic acid probe for HPV genome sequences directly to Pap smears or cervical tissue biopsies. The probe is tagged with an enzyme as a marker. After hybridization, cytochemical stains are carried out to demonstrate whether any hybridization has occurred. Figure 8.9a shows a photomicrograph of a cervical biopsy demonstrating in situ hybridization to a probe

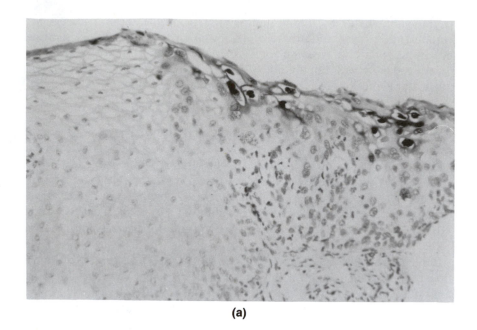

(a)

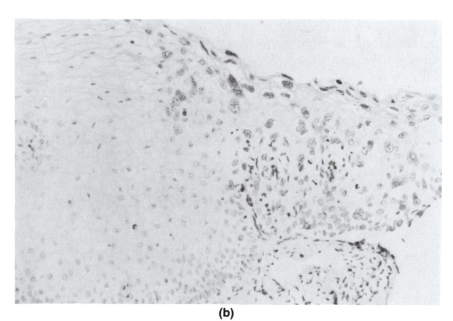

(b)

FIGURE 8.9. A photomicrograph of a cervical biopsy with dysplasia after in situ hybridization for HPV strain 33 (a) and strain 6 (b). The positive hybridization for strain 33 is associated with a more rapid clinical progression to a malignant tumor. (These photographs are from the laboratory of Dr. Katherine Pryzwansky, University of North Carolina, Chapel Hill.)

for HPV strain 33. The second panel in Figure 8.9b shows a section of the same tissue probed for HPV strain 6, which reveals a negative reaction.

Dot-blot hybridization detects HPV in DNA extracted from cells aspirated or scraped from the cervix. The sample does not have to be processed in a manner that preserves cellular detail for microscopic examination, simplifying collection in a nonmedical setting. The sample is chemically processed to lyse cells and expose the DNA. The lysed material is deposited by suction as a dot on a nylon membrane. The charged surface of the membrane preferentially binds DNA. A probe is hybridized to the membrane and a cytochemical color development reaction is carried out. The techniques for in situ and for dot-blot hybridization are well worked out, and both of these methods are available as commercial kits for use in clinical laboratories.

The incidence of routinely screened healthy females with HPV infection as detected by in situ hybridization or dot-blot techniques varies from 15 to 30%. Type-specific DNA probes show that only certain strains of HPV are associated with progression to malignant disease. HPV 6 or 11 strains are generally associated with benign genital warts with very low probability of malignant progression. HPV strains 16 and 18, and to a lesser extent strains 31, 33, and 35, are associated with a high rate of progression to malignant disease. The most common strain seen in lesions that are invasive at the time of initial screening is HPV 16.

There is still a good deal of controversy surrounding the use of HPV testing for the management of patients with cervical dysplasia. Some groups now recommend that all women having Pap smears be screened for the presence of HPV infection. If HPV is found, strain-specific studies should be done. Patients with strains 16, 18, 31, 33, and 35 should be followed at close intervals by repeat smears, biopsy, and colposcopy. Other groups feel that this is a premature recommendation. They recommend that only patients with cervical dysplasia as evidenced on a Pap smear have HPV testing carried out.

As with other new molecular tests, the debate about the optimal technology and how to use the data continues. Large-scale clinical studies are under way that should answer these questions within a relatively short time. The large number of Pap smears analyzed in the United States suggests that additional technologies could help improve the cost-effectiveness of this important cancer-screening test. Pap smears have been demonstrated to be highly effective when carried out under a program of quality assurance for both collecting and interpreting the smears. Methods to automate the Pap smear by pattern recognition, microscopy, or flow cytometry have to date had only limited success. Molecular diagnostic techniques, particularly when automated, may provide a cost-effective quality screening method to supplement the Pap smear as a screening method for cervical cancer.

Future Directions

A goal of molecular diagnostic studies in cancer is to analyze DNA from a biopsy specimen in order to determine the biological grade of a tumor in a more precise fashion than by histopathology alone. The heterogeneity of most caners makes it very difficult to gauge the success of a specific form of treatment. If patients can be more precisely staged by molecular diagnostics, then new therapies can be more rapidly evaluated.

There is also the hope that understanding cancer at the molecular level will lead to entirely new forms of therapy. It may be possible to interfere with defective cellular process by antisense oligonucleotides, supplemental growth factors, or genetically engineered antibodies which destroy the tumor cell.

The application of recombinant DNA technology to the clinical care of cancer patients has just begun but will probably progress rapidly. For some diseases, we have already identified specific molecular events that have strong prognostic significance. Colon cancer was an excellent example of sequential molecular events in a multistep process. For lymphomas and leukemias, the discovery of gene rearrangement and translocation of oncogenes appear to be the molecular basis for the abnormal cell growth in these tumors. For most other forms of cancer our knowledge of molecular events is incomplete, but improving rapidly.

Bibliography

Ayscue LH, Ross DW, Ozer H, Rao K, Gulley ML, Dent GA (1990) *Bcr/abl* recombinant DNA analysis versus karyotype in the diagnosis and therapeutic monitoring of chronic myeloid leukemia. Am J Clin Pathol 94:404–409.

Bayever E, Iversen P (1994) Oligonucleotides in the treatment of leukemia. Hematol Oncol 12:9–14.

Burkitt D, O'Conor GT (1961) Malignant lymphoma in African children I. A clinical syndrome. Cancer 14:258–269.

Cole HM (ed) (1993) Diagnostic and therapeutic technology assessment: human papillomavirus DNA testing in the management of cervical neoplasia. JAMA 270:2975–2981.

Cossman J (ed) (1990) *Molecular Genetics in Cancer Diagnosis*. Elsevier, NY.

Cossman J, Zehnbauer B, Garrett CT, Smith LJ, Williams M, Jaffe ES, Hanson LO, Love J (1991) Gene rearrangements in the diagnosis of lymphoma/leukemia. Am J Clin Pathol 95:347–354.

Daley GQ, van Etten RA, Baltimore D (1990) Induction of chronic myelogenous leukemia in mice by the p210*bcr/abl* gene of the Philadelphia chromosome. Science 247:824–830.

de The H, Chomienne C, Lanotte M, Degos L, Dejean A (1990) The t(15;17) translocation of acute promyelocytic leukemia fuses the retinoic acid receptor alpha gene to a novel transcribed locus. Nature 347:558–561.

Fearon ER, Vogelstein B (1990) A genetic model for colorectal tumorigenesis. Cell 6(61):759–767.

Gulley ML, Dent GA, Ross DW (1992) Classification and staging of lymphoma by molecular genetics. Cancer (Suppl) 69:1600–1606.

Hamilton SR (1992) Molecular genetics of colorectal carcinoma. Cancer 70:1216–1221.

Harris CC, Hollstein M (1993) Clinical implications of the p53 tumor suppressor gene. N Engl J Med 329:1318–1327.

Heisterkamp N, Stam K, Groffen J, de Klein A, Grosveld G (1985) Structural organization of the *bcr* gene and its role in the Ph′ translocation. Nature 315: 758–761.

Hockenbery D, Nunez G, Milliman C, Schreiber RD, Korsmeyer SJ (1990) Bcl-2 is an inner mitochondrial membrane protein that blocks programmed cell death. Nature 348:334–336.

Hollstein M, Sidransky D, Vogelstein B, Harris CC (1991) p53 mutations in human cancers. Science 253:49–53.

Huang ME, Ye YI, Chen SR, Chai JR, Lu JX, Zhoa L, Gu LJ, Wang ZY (1988) Use of all *trans*-retinoic acid in the treatment of acute promyelocytic leukemia. Blood 72:567–571.

Jacquemier J, Penault F, Durst M, Parc P, Seradour B, Meynard P, Halfon P, Hassoun J (1990) Detection of five different human papillomavirus types in cervical lesions by in situ hybridization. Hum Pathol 21:911–917.

Jen J, Kim H, Piantadosi S, et al. (1994) Allelic loss of chromosome 18q and prognosis in colorectal cancer. N Engl J Med 331:213–221.

Kern SE, Fearon ER, Tersmette KWF, Enterline JP, Leppert M, Nakamura Y, White R, Vogelstein B, Hamilton S (1989) Allelic loss in colorectal carcinoma. JAMA 261(21):3099–3103.

Lane DP (1994) p53 and human cancers. Br Med Bull 50:582–599.

Longo L, Pandolfi PP, Biondi A, Rambaldi A, Mencarelli A, Lo Coco F, Diverio D, Pegoraro L, Avanzi G, Tabilio A, Zangrilli D, Alcalay M, Donti E, Grignani F, Pelicci PG (1990) Rearrangements and aberrant expression of the retinoic acid receptor alpha gene in acute promyelocytic leukemias. J Exp Med 172:1571–1575.

Lyons J, private communication; see also Lyons J, Landis CA, Harsh G, et al. (1990) Two G protein oncogenes in human endocrine tumors. Science 249:655–659.

McDonnell TJ, Korsmeyer SJ (1991) Progression from lymphoid hyperplasia to high-grade malignant lymphoma in mice transgenic for the t(14;18). Nature 349: 254–256.

National Advisory Council for Human Genome Research (1994) Statement on use of DNA testing for presymptomatic identification of cancer risk. JAMA 271:785.

Ross DW, Brunning RD, Kantarjian HM, Koeffler HP, Ozer H (1993) A proposed staging system for chronic myeloid leukemia, Cancer 71:3788–3791.

Shattuck-Eidens D, McClure H, Simard J, et al. (1995) A collaborative survey of 80 mutations in the brca1 breast and ovarian cancer susceptibility gene. JAMA 273: 535–541.

Slamon DJ, Godolphin W, Jones LA, et al. (1989) Studies of the HER-2/neu proto-oncogene in human breast and ovarian cancer. Science 244:707–712.

Stanbridge EF, Nowell PC (1990) Origins of human cancer revisited. Cell 63:867–874.

zur Hausen H (1987) Papillomaviruses in human cancer. Cancer 59:1692–1696.

CHAPTER 9

Environmental Medicine

Environmental Pathogens

An environmental pathogen may be defined as a physical, chemical, or biological agent that, when present in the environment, causes an increased risk of disease. Environmental pathogens are usually present at only very low levels and the diseases that they cause may go unnoticed as only a small increase in disease incidence above "natural" background levels. Environmental pathogens have very different effects depending on their concentration. At a high level, toxic chemicals cause disease by direct damage to tissues. Such toxins are of course a major environmental concern. Their presence in the environment is a serious but avoidable event. Our major concern will be with environmental pathogens present at much lower doses than a direct toxic level. Low-level environmental pathogens cause disease mostly by damaging DNA. The effect of low-level environmental pathogens on other molecules is usually insignificant. For example, the ultraviolet component of sunlight causes slow deterioration of subcutaneous connective tissue, but this does not cause significant disability. When sunlight damages DNA, however, the result may be a squamous-cell carcinoma. The real danger of an environmental pathogen is a mutagenic event resulting from damage to DNA. The accumulation of mutagenic events has the potential to transform a cell into a cancer. It is this biological amplification of a single base-pair mutation resulting in a large malignant tumor that makes environmental pathogens a health risk at very low doses.

Since one of the major health risks of environmental pathogens at very low dose is cancer, these agents are also called carcinogens. There is a problem with this terminology, because the term *carcinogen* is also used to denote any agent that causes cancer in an experimental system at any dose. When a chemical is branded a carcinogen because of its effect at high dose, we have a tendency to avoid it even at very low dose. A carcinogen only becomes an environmental pathogen when it is a potential source of disease due to its presence at a low level in the environment.

Environmental pathogens, in addition to their role in causing cancer, are implicated in mental retardation and fetal malformation. Relatively little is known about the effects of most environmental pathogens on the central nervous system. The toxic effects of heavy metals such as lead, or of organic solvents such as alcohol, are greatest on the developing brain in utero and for infants. The molecular mechanisms of the effects of toxins on the central nervous system are, for the most part, unknown. However, environmental pathogens probably cause fetal malformations acting similarly to the way they do in cancer. Fetal growth is a complex interplay of cell division and differentiation. Many genes critical to fetal development are expressed only during short periods within the 9-month human gestation. Any interference with gene expression or DNA metabolism is a potential source of fetal malformation.

The most controversial issue in environmental medicine is the question of what constitutes a medically acceptable lower threshold for exposure to an environmental pathogen. If an agent is a carcinogen, why should we be exposed to it at all? The debate has raged in science, in medicine, and in the popular press. The federal government and various states have regulated environmental pathogens based on very little data, but in response to real fears. The best-studied environmental pathogen is ionizing radiation from x-rays and nuclear sources. To get a more in-depth understanding for the problem of environmental pathogens, I will explore the health risks of radiation in some detail.

Radiation

Ionizing radiation causes damage to molecules by direct physical interaction with molecular structure. The energy, dose, type, and length of exposure to radiation determine the amount of damage to tissues. An electrically charged particle, such as a high-energy electron beam, interacts strongly with molecules and does a lot of damage but does not penetrate deeply. An uncharged particle such as an x-ray photon penetrates more deeply and spreads its damage over a longer track. A strong basis in theoretical physics and a great deal of experimental data have created quantitative measures of the amount of damage to biological systems for any exposure. The standard measures for radiation exposure are the sievert (Sv) and gray (Gy). Formerly the units of radiation exposure were the rad or rem. Now 1 Sv has been defined as equal to 100 rem, and 1 Gy equals 100 rad. Further refinements for determining the amount of damage resulting from a given exposure include other measures such as the LET (linear energy transfer). For example, low-energy electrons of 0.1 MeV (millions of electron volts) penetrate less that 1 mm into the skin and do not damage deeper tissues. High-energy electrons of 10 MeV have an average penetration of 5 cm into biological tissues. If the same exposure of 1 Sv were to be from x-rays rather than electrons, the effect would again be very different. The low-energy x-rays would cause a significant skin burn; the high-energy x-rays would burn a deep organ such as the liver. When the dose, energy, and type of particle are known, the amount of damage that will be done to the biological system by radiation can be estimated. The amount of radiation exposure necessary to cause serious direct damage to an organism is quite high. Based on animal experiments and a review of human deaths following nuclear accidents, health physicists estimate that a dose of 4 to 5 Sv of high-energy radiation whole-body exposure will kill 50% of exposed humans. This level of environmental exposure is achievable only by close proximity to high-level nuclear isotopes. Exposures such as this would occur after a nuclear explosion or very near a major reactor disaster. Looking at the data available in the era after the Second World War, scientists concluded that by analogy humans would be entirely safe as long as they were not exposed to more than 0.15 Sv. Table 9.1a shows the permissible exposures limits for humans established then by various official panels and the continuing evolution of these limits to this day (Sinclair, 1987).

The much lower level of possible radiation exposure occurring from the use of medical x-rays, natural background cosmic radiation, radiation from natural radioactive elements, and from low-level nuclear wastes in the environment poses a very different problem (Upton, 1982). The average annual exposure due to these causes is less than 0.0002 Sv. Figure 9.1 shows the various sources of this annual exposure to the general public. Note that most of this low-level exposure is from "natural" sources, half being from radon gas in the environment. Only a small percent, about 15%, is added to the environment by human activities. The percent of humans who would be killed by the direct toxic effect of this typical annual exposure is zero. Radiation at this dose level does inconsequential damage to tissues and to all molecules except DNA. Thus the problem of low-level radiation exposure is entirely damage to DNA. Damage to DNA is also the major problem for low-level exposure to most other environmental pathogens.

Epidemiologists were the first to demonstrate an association between radiation exposure and increased cancer risk independent of any direct toxic effect. A study of dial painters of the 1920s was one of the first to demonstrate this risk. Thousands of workers in the New York area painted watch dials with radium so that the dial faces would glow in the dark. As a group, they were found to have a very high incidence of tissue breakdown in the mouth with osteomyelitis. Surprisingly some were also found to have aplastic anemia. Later, they were discovered to have a high incidence of cancers of the mouth, usually bone tumors of the mandible. Epidemiologists related this unexpected disease incidence to the workers' practice of licking the tips of their paint brushes to bring them to a fine point. This innocent act repeatedly deposited radium on the tongue. Radium is a naturally occurring radioactive element with a long half-life. The dial painters had accumulated large amounts of the radium in their tissues and were subject to continuous

Table 9.1a. Recommended limits for human exposure to radiation.

Maximum exposure[1]			
Occupational	Public	Year	Panel
0.3–0.6		1934	NCRP,[4] ICRP[5]
0.15		1949	NCRP, ICRP
0.05	0.010/30 year	1957	NCRP
0.05	0.005[3]	1971	NCRP

Table 9.1b. Lifetime risk of cancer from radiation exposure.

Risk[2]	Year	Panel
0.4%	1965	ICRP
0.6%	1970	Space Sciences Board
1–2%	1971	NCRP
1.0%	1977	ICRP
0.8%–2.2%	1980	BEIR III[6]
8.0%	1990	BEIR V

[1]Annual permissible exposure in sieverts (Sv).

[2]Lifetime risk of fatal malignancies per sievert (Sv). Thus if 1,000,000 people were exposed to 1 Sv; assuming the highest risk estimate of 8% issued in 1990, 80,000 would die of cancer attributed to radiation exposure. Of note, an additional 200,000 of this population would die of cancer due to other causes beside radiation. The 1-Sv exposure at an 8% risk level adds an incremental risk of 40% for cancer mortality.

[3]Average annual public dose is 0.0002 Sv, or about 4% of total permissible.

[4]NCRP — National Council on Radiation Protection.

[5]ICRP — International Council on Radiation Protection.

[6]BEIR — Panel on Biological Effects of Ionizing Radiation.

exposure. Madame Curie, who discovered the radioactive properties of radium, died of anemia secondary to malignant changes in her bone marrow from handling radioactive materials.

Threshold Dose

Once the cancer risk in radiation exposure was believed, the short-term toxicity studies performed in experimental animals were no longer valid as a means of determining safe exposure limits. Most of the studies of the 1950s looked only at acute effects of radiation damage. This led to an

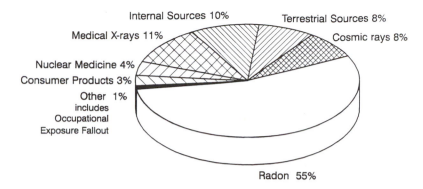

FIGURE 9.1. The various sources of radiation exposure that make up the total background exposure to the general public are shown in this pie chart.

incorrect assessment of the safe lower limit for exposure. Subsequently, scientists continued their detailed review of all human exposure to radiation to assess the cancer risk. Figure 9.2 shows data from a retrospective study that reviews the incidence of thyroid tumors in children who had their head and necks irradiated with x-rays (Favus et al., 1976; Shore et al., 1984).

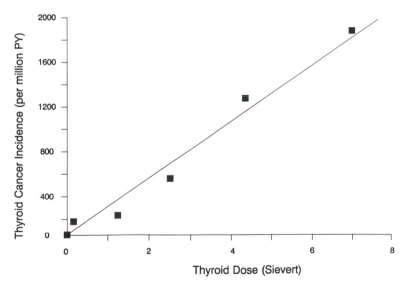

FIGURE 9.2. The incidence of thyroid cancer following thymic irradiation in children is plotted as a function of estimated dose to the thyroid in this retrospective analysis of these clinical data. The cancer incidence is given in number of cases per million patient years (PY) of risk.

This was a medical procedure practiced in the 1950s and early 1960s as an alternative to tonsillectomy. A substantial incidence of thyroid tumors was noted decades later. Other studies of humans exposed to radiation include Japanese survivors of the atomic bomb, women who have received breast irradiation, and people exposed to radioactive fallout from early above-ground nuclear weapons tests. Vehement debate over the question of what a safe limit for radiation exposure is has continued for more than 40 years. The question of how to establish a lower limit for exposure is central to the problem of environmental medicine. What we have learned for radiation is likely similar to what will be the case for a threshold for any environmental pathogen. Simply put, the question asks, "Is there a lower limit on radiation exposure below which essentially no harm occurs to the individual?" In 1990, the consensus on this question was no. The debate on lower limits is reflected in the steady downward trend in the permissible levels given in Table 9.1a.

How to establish a safe lower limit depends not only on accumulating data but on how to interpret the data. Figure 9.3 demonstrates data that are representative of animal experiments designed to estimate the cancer risk due to low-level radiation exposure (Ullrich, 1984). The incidence of malignant ovarian tumors in mice exposed to low levels of neutron radiation is plotted for various doses. As the dose decreases, the incidence of tumors decreases. The problem is that at the lowest exposure doses, where we are most interested in the cancer incidence, the data are the most imprecise. Several possible curves may be drawn through the measured data points. Each curve implies an important hypothesis about the effects of low-level radiation. A linear curve beginning at the origin is called the "linear no-threshold hypothesis." This hypothesis assumes that only at zero dose

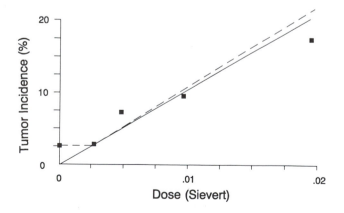

FIGURE 9.3. The incidence of ovarian tumors in mice after exposure to neutrons is plotted and analyzed to determine if there is a linear-no-threshold or linear-with-a-threshold relationship between exposure and cancer risk.

(which is not achievable because of background radiation) is a zero inci-dence of radiation-induced cancer achieved. The solid line in Figure 9.3 assumes a linear no-threshold relationship between dose and tumor inci-dence. Proponents of this curve, who are now in the majority, believe that it fits the data as well as any of the other models. And more importantly, a linear no-threshold hypothesis constitutes the safest assumption regarding low-level radiation exposure because it provides a worst-case scenario for tumor incidence at the lowest doses. A second hypothesis, which was held by a majority for many years until the mid-1980s, assumes a linear increase in tumor incidence with increasing radiation dose, but with a threshold dose below which no further decrease in cancer is achieved by decreasing exposure. This model is called the linear threshold hypothesis. Other names for this hypothesis are the hockey stick or quadratic hypothesis based on the shape of the dotted curve shown in Figure 9.3. This hypothesis was used in the calculation of safe radiation exposures for many years.

It is very difficult to get data on the cancer incidence for doses just above background level. The proponents of the linear threshold model point out that at very low doses most of the damage to DNA is repaired by the cell. The no-threshold proponents maintain, however, that if very large numbers of people are exposed to low levels of radiation, increased numbers of cancers are to be expected. Their motto is, "An infinitesimal risk applied to a very large number of exposed individuals will result in real deaths." It is not possible to achieve zero-dose radiation exposure since everyone is ex-posed to background radiation. It becomes increasingly more costly to limit exposure as we get closer to the background level. For example, the background dose in Denver, Colorado, is twice that of a similar city at sea level. This higher background is due to more cosmic rays penetrating the thin atmosphere above Denver. The expected cancer fatalities associated with low-level radiation exposure are given in Table 9.1b. Note that cancer risk has been revised upward innumerable times as our assessment of the risk for radiation-induced carcinogenesis at very low doses has increased. It is necessary to put the cancer risks from low-level radiation shown in Table 9.1b in perspective. The risk of a fatal cancer resulting from 1 year's expo-sure to background radiation is equivalent to the risk of a fatal motor vehicle accident on a 500-mile trip. The public's perception of the health risks of low-level radiation has always been higher than its concern for other risks (Slovic, 1987). This undoubtedly relates to the fear of nuclear war or accidents, which can result in high-level radiation exposure with cataclysmal health effects.

Chemicals

The problem of chemicals as environmental pathogens is an order of magni-tude more complex than the situation for radiation. The study and debate surrounding radiation as an environmental pathogen span 40 years and have been fueled by a massive research effort funded by the Atomic Energy

Commission (now part of the Department of Energy). The controversy about radiation is not over, but some agreement has been reached. Now there remain thousands of chemical environmental pathogens to consider, none of which has been studied a fraction as well as radiation. One hopes that some of what we have learned about radiation will apply at least in principle to other agents. It remains, however, very difficult to determine the danger of the huge spectrum of natural and artificial chemicals in the environment with respect to their ability to induce cancer. Chemicals vary in the way they enter the body, in the way they are metabolized, where they become concentrated, and how they interact. For most chemicals we have no simple method for measuring exposure dose, nor do we have a good quantitative estimate of the level of risk. Table 9.2 lists an upper estimate of the risk for exposure to a few air-borne carcinogens. These risk estimates assume a roughly continuous exposure over a long period of time at an exposure of 1 $\mu g/m^3$. These estimates are approximate, but they demonstrate that risks between various chemicals vary by orders of magnitude. Long-term exposure to dioxin carries a potential certainty for the development of cancer (risk = 1); the same level of exposure to formaldehyde carries a one-in-a-million risk.

The health effects of chemical environmental pathogens constitute a problem of immense scientific and social complexity. To begin, we need quantitative tests of the carcinogenic potential for chemicals. This has not been an easy goal to achieve. Methods for determining the cancer-causing potential of chemicals include animal assays, in vitro mutagenesis tests, and epidemiological studies. Animal testing has been one of the mainstays of carcinogenesis research. A suspected carcinogen is given to a number of mice or rats under appropriately monitored conditions. After a period of time, the incidence of tumors in the animals is compared with an unexposed control group. However, animal testing is an expensive, controversial, and time-consuming method for screening potential carcinogens. In 1975, Bruce Ames of the University of California at Berkeley introduced a relatively

Table 9.2. Upper-limit risks for airborne carcinogens.[1]

Chemical	Risk estimate
Dioxin	1
N-nitroso-N-ethylurea (NEU)	0.01
Diethylnitrosamine (DEN)	0.002
Ethylene oxide	0.0002
Formaldehyde	0.00005
Benzene	0.000007

[1]Excess lifetime risk of cancer associated with breathing 1 $\mu g/m^3$ over a 70-year life span (from NCRP, 1989).

simple in vitro test to screen for chemical carcinogens (Ames et al., 1975). The Ames test measures the mutagenic potential of chemicals. A major assumption of the test is that mutagenesis in bacteria is closely related to carcinogenesis in animals. In the Ames mutagenesis test, the chemical under question is applied to a petri plate that lacks an essential growth nutrient. The plate is then seeded with *Salmonella* bacteria. If a mutation has occurred that permits a bacterium to grow independently of the previously essential nutrient substrate, a colony will be formed on the plate. Bacteria that are not mutated will fail to grow. Of the approximately 100 million bacteria present on the original plate, even one mutant colony can be detected that would indicate a mutation rate of 10^{-8}. The assay is sensitive, quantitative, and easy to carry out. Studies of a new chemical that might take several years to carry out in animals can be completed within months by using the Ames test.

After the advent of the Ames mutagenesis test, thousands of compounds were studied. Mutagenic potential was shown for a large number of substances present both naturally and artificially in our environment, particularly in food. These data led to a debate on the relevance of the Ames mutagenesis assay in carcinogenesis. Many government agencies began regulating exposure to compounds shown to be mutagenic at doses that humans might encounter. The data from the Ames mutagenesis assay and other similar assays led to environmental regulations that first received widespread implementation in California.

However, much controversy surrounds the use of data on the carcinogenic potential of chemicals. Ames feels that too much emphasis has been placed on man-made chemicals, when in fact he claims that we ingest 10,000 times more natural pesticides than man-made ones (Ames et al., 1987). From the list of just a few carcinogens noted in Table 9.2, Ames points out that benzene, formaldehyde, and diethylnitrosamine are all found "naturally" in foods. However, there is by no means agreement on the question of natural vs. artificial carcinogens, nor is the worth of the Ames test certain. Weinstein (1991) has argued that the Ames mutagenesis assay only tests the effects of chemicals on dividing cells. He points out that cancer is a multistep process, and the Ames assay only looks at one step.

The debate regarding threshold vs. no threshold for radiation-induced cancers is repeated for chemical environmental pathogens. For each agent, the question once again is asked, "Is there a lower level below which no risk occurs from exposure, or does the risk continually decrease with dose but never go to zero?" As for radiation carcinogenesis, the consensus opinion is that there is no lower threshold for most cancer-causing chemicals. However, we are uncertain about the quantitative relationship between dose and risk for most chemical carcinogens such as dioxin or biological agents such as aflatoxins produced by fungi. Furthermore, our ability to measure very low levels of these chemicals is much more limited than for radiation. A Geiger counter, which costs several hundred dollars, can measure radiation exposure at any dose range from background to very high levels. To screen soil, water, and air samples for chemical environmental pathogens at very low doses is both expensive and, in some instances, not technologically possible.

Molecular Testing of Environmental Pathogens

Recombinant DNA technology may be the tool that provides the breakthrough necessary to resolve the complex issues of environmental pathogen exposure, threshold, and risk. Rather than measure indirect effects of potential pathogens in vitro or in animal models, recombinant DNA technology makes it possible to measure the end result, damage to DNA in humans. People with different environment exposures can be compared to look for factors that increase the damage to DNA. The measurement of damage to DNA can include all mutations. Alternatively, damage can be assessed at specific "hot spots" such as within oncogenes.

Recombinant DNA methods can assess such changes as methylation of nucleotides, presence of DNA adducts (binding of foreign substances to the DNA molecule), rate of somatic-cell mutations, and chromosome breaks. The measurement of the effects of environmental pathogens in humans by using molecular diagnostic techniques is just beginning. As an example of the kinds of studies under way, consider the search for the effects of organic solvent exposure as a stimulus for leukemia. In patients with acute myeloid leukemia, a higher incidence of *ras* oncogene mutation is seen in persons with a history of solvent exposure. The *ras* oncogene when mutated has a high potential for inducing malignancies and may contribute to the evolution of leukemia. Epidemiologic studies are quite important in that they demonstrate associations, if not cause. Much larger scale prospective studies are needed to monitor subjects in good health, watching for a correlation between DNA damage and the development of cancer. More research into the mechanisms of environmental-pathogen-induced DNA damage will be necessary as a precursor to large screening studies.

The technology for measuring DNA damage has not yet reached the point where it is feasible as a routine part of health screening. Currently assessment of DNA damage in humans is possible in research laboratories by using somewhat laborious techniques. Automation of DNA testing will be necessary to bring these methods to a scale necessary for use in screening large numbers of people. Nevertheless, a number of industries have expressed a desire for tests that will allow them to screen and monitor the effects of potentially dangerous environmental exposures on their workers. The impetus for the development of clinical tests of DNA damage is very strong, and I anticipate that in the near future such tests will be available. A test that measures the amount of damage to the DNA in a sample of peripheral blood lymphocytes would allow for an assessment of past exposure to environmental pathogens. With this information, physicians could give more specific advice on the relative risk of future exposure. For example, it is already possible to quantitate the smoking history of a patient by measuring breaks and abnormal methylation patterns in a sample of the subject's DNA.

In the near future, physicians will be faced with a bewildering array of new DNA-based diagnostic tests. There will be DNA-based tests for specific diseases as well as tests that are designed as general measures of health risks due to environmental pathogens. If these tests are not used by physicians as part of the overall health-care maintenance of patients, it is likely that a nonmedical environmental health practice will arise. The limited experience with radon testing of homes has already demonstrated this issue. Radon is a risk factor for lung cancer and has been very well publicized for the last 5 years. Numerous kits are available in drug stores, supermarkets, and through the mail for radon testing. Patients do not usually bring the results of radon testing in their homes to physicians. They instead consult public health agencies or the manufacturers of the kits for recommendations regarding their health risk and possible removal of radon exposure. Physicians could better advise patients of the relative risk of radon exposure by considering not only radon levels in homes but all the other health patterns of the subject. Radon exposure in smokers, for instance, is a significantly greater risk than in nonsmokers. It remains an important challenge for physicians to learn and adapt to this virtually new branch of medicine.

Bibliography

Adelstein SJ (1987) Uncertainty and relative risks of radiation exposure. JAMA 258(5):655–658.

Ames BN, Magaw R, Gold LS (1987) Ranking possible carcinogenic hazards. Science 236:271–279.

Ames BN, McCann J, Yamasaki E (1975) Method for detecting carcinogens and mutagens with the *Salmonella*/mammalian microsome test. Mutat Res 31:347–364.

Favus MJ, Schneider AB, Stachura ME, Arnold JE, Yun Ryo U, Pinsky SM, Colman M, Arnold MJ, Frohman LA (1976) Thyroid cancer as a late consequence of head and neck irradiation. N Engl J Med 294(19):1019–1025.

Hendee WR (1992) Estimation of radiation risks. JAMA 268:620–624.

Loken MK (1987) Physicians' obligations in radiation issues. JAMA 258(5):673–676.

Macklis RM (1990) Radithor and the era of mild radium therapy. JAMA 264(5):614–623.

National Council on Radiation Protection and Measurements (1989) Comparative Carcinogenicity of Ionizing Radiation and Chemicals, Report No. 96. Bethesda, MD, p 125.

Shore RE, Woodward ED, Hempelman LH (1984) Radiation induced thyroid cancer. Boice JD, Fraumeni JF (eds) *Radiation Carcinogenesis*. Raven Press, NY, pp 131–138.

Sinclair W (1987) Risk, research and radiation protection. Radiat Res 112:191–216.

Slovic P (1987) Perception of risk. Science 236:280–285.

Ullrich RL (1984) Tumor induction in BALB/c mice after fractionated or protracted exposures to fission-spectrum neutrons. Radiat Res 97:587–597.

Upton AC (1982) The biological effects of low-level ionizing radiation. Sci Am 246(2):41–50.

Weinstein IB (1991) Mitogenesis is only one factor in carcinogenesis. Science 251: 387–388.

Index